HOMEGROWN
HONEY BEES

HOMEGROWN HONEY BEES

An Absolute Beginner's Guide to
BEEKEEPING Your First Year, from Hiving to Honey Harvest

ALETHEA MORRISON
Photography by MARS VILAUBI

Storey Publishing

*The mission of Storey Publishing is to serve our customers by
publishing practical information that encourages
personal independence in harmony with the environment.*

Edited by Deborah Burns
Art direction and book design by Alethea Morrison

Cover photography by © Mars Vilaubi
Interior photography by © Mars Vilaubi, except as noted on page 159
Illustrations by © David Wysotski, pages 23 and 30
Chapter opener and Urban Beekeeping illustrations by © Gilbert Ford

Indexed by Samantha Miller

The information in this book is true and complete
to the best of our knowledge. All recommendations
are made without guarantee on the part of the author
or Storey Publishing. The author and publisher dis-
claim any liability in connection with the use of this
information.

Storey books are available for special premium
and promotional uses and for customized editions. For
further information, please call 1-800-793-9396.

Storey Publishing
210 MASS MoCA Way
North Adams, MA 01247
www.storey.com

Printed in China by Toppan Leefung Printing Ltd.
10 9 8 7 6 5 4 3 2 1

Library of Congress Cataloging-in-Publication Data

Morrison, Alethea.
Homegrown honey bees / by Alethea Morrison ; photography by Mars Vilaubi.
    p. cm.
Includes bibliographical references and index.
ISBN 978-1-60342-994-8 (pbk. : alk. paper)
1. Bee culture—Handbooks, manuals, etc. 2. Honey. I. Title.
SF523.M6875 2013
638'.1—dc23

CONTENTS

Thanks So Much

Our bee mentor, Tony Pisano, has kept many more hives than we have for many more years. Beekeeping is his passion. He seems to be everywhere at once in our community, proselytizing to people who don't yet keep bees, educating and supporting those who do. Thank you, Tony, for collaborating on this book with us.

Special thanks also to Michael Palmer and Randy Oliver for reaffirming our belief that beekeepers are some of the most generous people in the world. These excellent, expert beekeepers donated their valuable time to give us helpful feedback.

We dedicate this book to Wangari Maathai for being such a deep well of inspiration. We carry only single drops of water to a raging fire, but we are doing the best that we can.

PREFACE

There are days in your life you never forget. For me it was my wedding day. The birth of our son. The day my first chicken laid her first egg.

The last one may not seem like a biggie, but it was epic. My family and I had taken our chickens home when they were one day old and coddled them in our house until they were big and smelly enough to have to move outside. Then we cosseted them some more until that extraordinary November day when we found one gloriously beautiful egg in the nesting box. I was so proud, you would think I had laid it myself.

My chickens' eggs taste better, and not just because they couldn't be any fresher. There's a joy to DIY, whether it's growing your own food, building a bookcase, reupholstering a chair, or knitting a scarf. I love this, I said to myself. What can I do to keep my DIY buzz on? Why, keep bees, of course!

My husband, Mars, is a photographer and an obsessive documentarian. During the first couple of years that we kept bees, he photographed how to install a new package of bees, light a smoker, inspect a hive, manage pests, harvest honey, and almost everything in between. We understood what beginners needed to know because we were learning ourselves at the same time. Nothing was second nature, so nothing was overlooked.

I've approached the writing of this book in the same way, spelling out every step as clearly and concisely as possible for the absolute beginner. I'm also sharing our personal experiences and those of other beekeepers I've met along the way, because this journey is much livelier than basic facts and figures. You'll laugh, you'll cry — if not at my stories, then at your own as you become a beekeeper, too. Beekeeping is fun, fascinating, and heartbreaking. Kind of like life.

chapter 1

WHY BEES?

Sweet Alchemy

While many species of bees are pollinators, as are butterflies and other insects, honey bees are especially valuable to agriculture, because they let us keep them in hives. Commercial beekeepers transport hives from crop to crop, traveling hundreds or thousands of miles throughout the year. Honey bees' economic value as pollinators actually far surpasses the value of their honey crop, but it's the honey crop that motivates most backyard beekeepers.

Like wine or coffee, honey doesn't have one flavor. Nectar from different flowers creates honey with different essences. If you hope to produce honey from just one type of flower, such as clover, orange blossom, tupelo, buckwheat, or dandelion, your beehives must be in an area where that flower dominates, and as soon as those blossoms pass, you must harvest the honey or move the hives.

While monofloral honey is delicious and wonderful, wildflower honey is nothing short of astonishing. It may be concocted from a medley of different nectars, but each bottle is unique and special to the moment in time and place that the bees made it. I harvest honey at the same time as a beekeeping friend who lives less than two miles away, and our crops are totally different in color and flavor. Honey is a distillation of the miniscule spot on the planet that you and your bees inhabit: the transmutation of the land into liquid gold.

Aristotle surmised that bees collected honey from rainbows. He may not have gotten the science right, but he was spot on with the poetry of its flavor.

Honey may be the star of the show, but beeswax is another valuable product of the hive and is often used to make candles and skin care products. Beeswax candles have a gently sweet scent and don't smoke when they burn. In cosmetics and creams, the wax protects and softens the skin.

Perfect Purity

IT'S A SMALL WONDER that civilizations have often regarded honey as holy. It is so perfect and clean that it crystallizes but never spoils. Honey that your bees make today you could still eat three thousand years from now if you happened to be immortal. Unspoiled honey has frequently been found in ancient Egyptian tombs, though I doubt there have been many takers to taste it, since the Egyptians often used honey to preserve bodies.

Bee Diary

Our bee club has an annual banquet. Each person brings a 1-pound jar of honey, and we have a honey-tasting contest, with prizes for best color and best flavor. As we roam around with our dipping sticks, we taste dozens of samples, each remarkably unique. Honey has a terroir every bit as complex as wine's.

Power to the People

Across the political spectrum today, there is a sense of helplessness in the face of forces beyond our personal control, whether it's climate change, terrorism, economic insecurity, unhealthy and unsafe food, or government intrusions on our liberty. If any of these perils makes you want to become more self-sufficient, you are not alone. There are more reasons to keep bees than delighting your taste buds.

» **Worried about environmental, economic, and health consequences** of an industrialized food system? Harvest your own local sweetener!

» **Want to get closer to nature?** Bees are more like wild animals than domesticated livestock, and beekeepers have the privilege to observe their lives and behaviors up close. Beekeepers also become highly attuned to the local environment, as the success of the honey harvest depends so much on the weather and the health of nectar- and pollen-producing plants.

» **Skeptical of pharmaceuticals** and interested in natural health remedies? Have ready access to raw honey, bee pollen, and other products of the hive!

» **Planning an independent food supply** in case all hell breaks loose? Ensure that you'll keep life on the homestead sweet!

Keeping bees is an empowering action, however small, that makes you part of the solution.

Raw Power

COMMERCIAL HONEY is usually heated and filtered. Honey that is not treated is known as raw, and many believe that it offers myriad health benefits that are lacking or diminished in processed honey, such as:

» Boosting the immune system, with as little as a teaspoon a day
» Acting as an antibiotic to care for wounds — from minor cuts and scrapes to serious burns
» Providing a source of antioxidants
» Aiding digestion and other gastrointestinal functions
» Moisturizing the skin
» Soothing sore throats and treating coughs
» Managing allergies
» Curing hangovers

There are also claims regarding the health benefits of pollen, propolis, royal jelly, and even bee venom. Applying bee stings medicinally, known as apitherapy, is said to relieve arthritis.

(Note: Some of these claims are not yet proven to the standards of evidence-based allopathic medicine.)

Power to the Pollinators

Speaking of troubled times, the world is in a pollination crisis. If you haven't spent the last decade in a subterranean hermitage, you've probably heard of colony collapse disorder (CCD), the phenomenon in which bees mysteriously, suddenly, and completely disappear from their hives. Without definitive evidence about the cause of CCD, the United States has not developed an effective policy to manage it. Bees are dying at such a rapid clip that there are no longer enough hives to pollinate the nation's commercial crops, one-third of which rely on honey bees. Now is a great time to take up beekeeping and help protect our future, one bee at a time.

SOME CROPS THAT HONEY BEES POLLINATE

Almonds

Apples

Blackberries

Blueberries

Cantaloupes

Cherries

Cranberries

Cucumbers

Peaches

Pears

Raspberries

Soybeans

Strawberries

Watermelons

Colony collapse disorder is less common in small-scale beeyards than it is in large-scale commercial operations.

Body Language

AS MATA HARI SAID, "The dance is a poem of which each movement is a word." Though bees don't have a spoken language, they speak volumes through dances. Rather than thousands of bees haphazardly seeking food, when one finds success she returns to the hive and begins vibrating her wings and swinging around in a series of circles or figure eights to communicate the direction, distance, quantity, and quality of a food source.

One dance movement is called the *waggle*. The angle of the waggle is the angle the bees must fly in relation to the sun to reach the forage. When forage is close, the style of the dance indicates distance. For longer distances, the number of waggles instructs how far to fly.

Curiouser and Curiouser

All that do-gooding and you get to have a madcap hobby, too! When I first got bees, I felt as if I had fallen down the rabbit hole. Here was an entire complex civilization that was hidden from my view until I stuck my nose in to explore. Their language, behaviors, and rules are confusing to the novice, and frankly, the inhabitants of this alien world aren't particularly friendly. But if you are daring and inquisitive (some may say crazy) enough to spend time with untamable creatures, bees are an experience that won't fail to excite.

Welcome to a wild, wonderful adventure.

"But I don't want to go among mad people," Alice remarked.

"Oh, you can't help that," said the Cat: "we're all mad here. I'm mad. You're mad."

"How do you know I'm mad?" said Alice.

"You must be," said the Cat, "or you wouldn't have come here."

— **LEWIS CARROLL**, *Alice's Adventures in Wonderland*

chapter 2

WHAT AM I GETTING INTO?

Time and Money

For a garden to produce a robust yield of delicious food instead of a wasteland of dead plants, it requires routine maintenance, specific tasks done at specific times, and your consistent awareness of how it is developing. Beekeeping is the same way. As a very approximate estimate, plan on spending an hour a week during the bees' active season if you have just a couple of hives. When you harvest honey, you'll be busier than that, and if your hives rest in the winter, you'll rest then, too.

Beekeeping's start-up costs are significant enough to merit budgeting for, but once you're underway it is fairly inexpensive. You'll spend more than you would for a video game console and way more than for some knitting needles and a few skeins of yarn, but with those hobbies and most others, you must continually feed the habit. Beekeeping is relatively self-sustaining.

Read chapters 4 and 5 about prepping a bee yard, buying bees, and all the equipment you need so you can put together a shopping list. Get catalogs from some beekeeping suppliers listed in resources on page 153 to have a solid idea of how many Ben Franklins you'll need to invest.

If money is tight, consider top-bar beekeeping, which demands a smaller outlay of cash (see page 70).

Drew and Ali Johnson

KALAMAZOO, MICHIGAN

When Drew and Ali bought their first home last year on a quarter of an acre, they immediately started gardening and raising chickens. Bees were a natural next step and a good fit for the small size of their homestead.

Drew first intended to get a typical Langstroth hive, but the low costs of top-bar beekeeping were too appealing to ignore. For an investment of $30 and a day's work, he set up his hive from plans he found on the Internet. His only other purchase was the bees themselves. No safety gear? No sweat. He just uses mosquito netting to protect his face and duct tape to keep the bees from crawling inside his clothes.

Education

Beekeeping isn't easy or simple, but fortunately, you don't have to go it alone. Every beekeeping book will offer a new and valuable insight, so read as much as you can. Recommended books are listed in the resources section.

Research and reading don't have to end with books and magazines. You can also:

Get online. Beesource.com is an active online community where people post information and have conversations about every beekeeping topic imaginable. It can be a great place to ask questions and get feedback.

Take a workshop. Many beginning beekeepers take workshops before setting up their first hives. Those can be a great introduction to beekeeping or can reinforce and help you absorb the information you're gathering. You can also ask questions and make connections with other people who keep bees or plan to. Local bee associations often offer these classes.

Join the club. Speaking of bee associations, it's important to reach out to one if there is a club in your area, even if you're not interested in a workshop. Beesource.com has a pretty thorough listing of local associations. Books and online forums are great, but there is no substitute for folks living nearby who can give you information about beekeeping in your specific area and help you out in person when you need it.

If you manage to score a bee mentor at an association meeting, I guarantee you will double the fun of your first year of beekeeping. Don't be shy. Beekeepers love what they do, and many of them are happy to share their wisdom with someone as inexperienced but enthusiastic as you! Truly, they only wish more people were interested.

Bee Diary

Our bee mentor, Tony, was remarkably generous with his time our first year of beekeeping. If I wasn't e-mailing him with a question, I was sharing some beekeeping joy or hardship. No one understands your little triumphs and tragedies better than a fellow beekeeper. And when the soup really hit the fan with one of our hives, he had our back. He also let us use all his honey harvesting equipment so we didn't have to spend so much cash the first year.

Amy Azzarito

BROOKLYN, NEW YORK

Amy is a reformed outlaw, or, rather, New York City is a reformed sheriff. Before the city overturned its ill-advised law against beekeeping in 2010, Amy had persuaded her friend Barry Rice to use his Brooklyn rooftop as a secret apiary and to go in with her on a hive. They took a class at the New York City Beekeepers Association and were gladly swept into the swelling tide of urban agriculturists.

Besides workshops like the one Amy and Barry took, the association offers area beekeepers a lot of support, such as monthly meetings, newsletters, guest speakers, group orders on package bees, loans of expensive equipment like extractors, and an accessible community of mentors. The association even matches up people who want to keep bees but don't have a space with bee-friendly folks who are willing to host a hive in their garden, balcony, or rooftop. Amy has found her strong network of fellow beekeepers to be a great resource as she navigates the newbie's inevitable pitfalls and pratfalls.

The Fear Factor

Are you wondering whether beekeepers get stung, how often, and how much it will hurt? The answers, in order, are: yes; it depends; and not too badly. If you're very risk-averse, you can wear the beekeeping equivalent of a hazmat suit that offers full-body protection. I've never met a beekeeper that hasn't been stung, though, so get used to the idea. I promise that it's really not so bad (unless you have a true systemic allergy).

How often you get stung depends on how much protection you wear, how skillfully you work the hive, and the temperament of your bees. For example, a beekeeper blundering nervously about will get stung more often than one moving slowly and calmly. Opening the hive when it's cool or overcast will also raise the bees' defenses. As you read further in this book, you'll learn how to work the hive sensitively and smartly so the bees are less likely to sting you.

Contrary to our culture's deep-rooted fear of stinging insects, honey bees really aren't out to get you, and they aren't likely to sting you or your neighbors as you go about your daily business. When you open up their house and start poking around where they raise their babies and store their food, that's another matter, and they will be on the alert at the very least. I'm sure you can relate.

ALLERGY ALERT

The vast majority of people experience only minor swelling, redness, and itching at the site of a bee sting. Even if the swelling is much more severe than average, you are not in any danger. An antihistamine like Benadryl can help relieve symptoms.

Some people, however, suffer anaphylactic shock, causing difficulty in breathing, which requires immediate medical attention. An epinephrine injector like the EpiPen is available with a doctor's prescription for emergency treatment.

Kamikaze Mission

IT MAY BE SMALL CONSOLATION in the heat of the moment, but a sting hurts the bee more than it hurts you. As she nails you and flies away, the stinger is torn from her body, and she dies with her entrails literally dripping out. You have to feel a little sorry for her, taking that kind of hit to protect her family.

A queen's stinger, on the other hand, is not barbed so she doesn't die when she stings.

Bee Diary

It's intense to open a hive and stand in a cloud of somewhat aggravated insects ready to put the hurt on you. Beekeeping is not for the faint of heart. It's like the extreme sport of animal husbandry.

Law and Order

You'd think with all the beauty and abundance bees bring to the landscape and pantry by pollinating flowers and crops, they would have a better reputation. Unfortunately, they are commonly feared and loathed as nuisances, and many municipalities regulate beekeeping or prohibit it outright. Find out what the rules are where you live, ideally from your local bee association. It can give you the lowdown much more quickly than can Byzantine government offices and websites.

With the serious decline of honey bee populations, and the media focusing on the threat to our food supply, now is a great time to advocate for lifting these bans. After many years of effort, activists in New York City successfully lobbied to legalize beekeeping in 2010, giving official sanction to an activity that many people were already doing surreptitiously throughout the most densely populated city in the nation.

GOOD BEEKEEPERS ARE GOOD NEIGHBORS

Even if you know how great bees are, your neighbors might not, so an education campaign is in order if your beeyard is your backyard. Someone could freak out if a couple of beehives unexpectedly pop up in the neighborhood one day. Discussing it with neighbors in advance will help them feel they have buy-in. Even if you have the legal right to keep bees, you'll have a lot more fun without confrontations or bad feelings.

Sharing your honey will make your hobby a lot more popular with friends and neighbors.

False ID

YELLOW JACKETS are the most aggressive of stinging insects and are very commonly misidentified as honey bees, since they both have striped bodies. Unfortunately, bees often take the rap for these wasps' pesky behavior. It pays to know the difference and spread the word to set the record straight.

Yellow jackets can be serious pests, nesting inside walls, for example.

HONEY BEE

body is furry ----→

bands of color range from deep gold to brown ----→

legs tuck against body when flying

YELLOW JACKET

body is shiny and smooth ----→

yellow bands are much brighter in color

legs hang down when flying

HONEY BEES	YELLOW JACKETS
Bee family (Apidae)	Wasp family (Vespidae)
Die after a single sting	Can sting many times without dying
Mildly defensive	More aggressively defensive
Bodies are covered with small hairs, giving them a slightly fuzzy appearance	Bodies are shiny and smooth
Eat nectar and pollen from flowers	Eat small insects, rotting fruit, and plant sap, and are common pests at cookouts, picnics, and garbage cans, trying to scavenge food with a meaty or sweet smell
Fly with their legs tucked in	Fly with their hind legs hanging down
Nest high off the ground	Nest in underground hollows or in the walls of buildings
Nest is wax	Nest is gray and papery

chapter 3

BEE LIFE 101

Journey to the Center of the Hive

Much of a colony's life is spent out of sight, and only beekeepers have the privilege of regularly observing this microsociety. While part of the fun of this adventure is getting a window into the secret lives of bees, what you learn is more than just a curiosity. Knowledge of the structure and cycles of the colony, as well as the bees' instincts and behaviors, is the foundation of good beekeeping. Every decision beekeepers make, from the equipment we buy to when we visit the hive and what we do when we're there, is based on our understanding of bee life.

Pollen can vary in color, from the obvious yellow and orange to more surprising shades of purple. When the pollen cell is almost full, a bee will top it off with a thin layer of honey as a preservative.

Here are the primary elements you will see inside a hive, aside from the equipment (also called *furniture*) you introduce as a beekeeper:

» **Brood and bees.** If the hive is in good health, all stages of bees are in the hive, from brood (eggs, larvae, and pupae) to adults.

» **Beeswax comb.** Bees secrete wax from glands on their abdomens. Then they chew the wax to make it pliable and sculpt it into interlocking hexagonal cells called *comb*. Inside these cells they raise their young and store their food.

These cells will soon be capped and the larvae will pupate.

» **Nectar, pollen, and honey.** Bees forage nectar and pollen from flowers and bring it back to the hive to store as food. They treat the nectar with an enzyme and fan their wings to evaporate its water content, curing it into honey. Since honey will absorb water out of the air and ferment if the moisture level gets too high, bees cap cured honey with wax. The cappings, much like jar lids, preserve the honey by keeping out the moisture.

The bees mix honey and pollen and ferment it into *bee bread,* a precious food for brood and the youngest bees.

Propolis is a sticky substance bees make from plant resin to fill small gaps and cracks.

» **Propolis.** Bees also forage plant sap and resin, which they treat with another enzyme to make propolis. Propolis is like a bee's duct tape, good for fixing just about everything, from sealing cracks and holes to strengthening wax comb. Propolis has antibacterial qualities, too, and the bees use it as an all-around cleaning product. Some colonies go overboard with the propolis. When they use it to glue hive parts together, it can be a real nuisance to the beekeeper.

You can identify honey cells by the cappings, which are flat and look either dry and opaque white or slightly wet and more transparent. Below, there is one cell of capped honey surrounded by cells full of either uncapped honey or uncured nectar.

Bee Space

BEES ARE STICKLERS. They want exactly ¼ to ⅜-inch passageways — called *bee space* — in their hive. Any smaller space they'll seal up with propolis. Any larger space they'll fill with comb.

Johann Dzierzon published his studies of bee space in 1838 and designed the first practical moveable-comb hive. In 1848 he introduced a new design that quickly gained popularity in Europe and the United States. Though L. L. Langstroth is often credited with discovering bee space, he didn't invent his own moveable-frame hive until 1852. Langstroth's design is the one that most beekeepers use today.

Before the popularization of such hives, many beekeepers were actually bee destroyers, ripping apart a nest to get to the honey. Langstroth's design hung comb in vertical sheets, placed just the right distance apart to maintain bee space, and used frames, so the beekeeper can remove the comb easily and have ready access to all the cells inside without damaging the structure of the hive.

This rogue comb on the frame below resembles honey bee nests in the wild, which they construct in wavelike sheets, separated by bee space.

The Citizenry

Honey bees show a level of social organization so extreme that individuality is meaningless. The instinct to survive and reproduce is held by the colony as a whole. No one bee in a social group competes against another for food or mating opportunities in the way mammals, birds, and even most other insects do. A honey bee's total devotion to its colony's well-being results in breathtaking acts of heroic self-sacrifice and brutal acts of premeditated murder.

In this insect opera there are three principal players: the queen, the drones, and the workers.

THE QUEEN MOTHER

A colony has one queen, and she has an outsize importance. Without her, a colony will wither and die. Since she is the only fertile lady in the hive, she lays all the eggs. And since she lays all the eggs, every bee born is her daughter or son.

They all carry her genetic material, so qualities like temperament, productivity, cleanliness, and disease resistance are inheritances that most of the bees in a colony will share. If the queen is a bad apple, she will definitely spoil the whole barrel.

The queen mates only a few days in her life, when she is about a week old. She flies from the hive and attracts cruising drones from other colonies, mating with them in succession. From those couplings she gets all the zillions of sperm cells that she will ever need over the course of her egg-laying years. Barring any disease or injury, she will reign supreme over her bees, laying up to two thousand eggs a day for two or more years.

Queens can be difficult to find in the hive, especially for new beekeepers. She runs from the light and tends to slip around to whatever side of the frame you are not inspecting. Some queen breeders mark the queen with a colored spot on her back, making her easier to see.

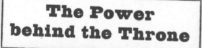

The Power behind the Throne

THE QUEEN RELEASES PHEROMONES that signal she is present, active, and healthy, holding the colony together and keeping it on task. Her powerful chemistry even inhibits the development of workers' ovaries so they can't lay eggs. But if she is sick, injured, or aging to a point of waning fertility, the bees are quickly wise to this and will raise a new queen to replace her. (See Supersedures, page 37.)

Note the egg she is about to lay in a cell!

queen

The queen is the longest bee in the colony. She has a smooth, hairless thorax; wider hips; and an abdomen that tapers gradually. (A virgin queen has a fuzzy thorax like other bees, but the constant grooming of her retinue quickly wears it off.)

The drone is large like a queen but looks fatter, with a thick waist and rounded abdomen. He has very big eyes that meet in the center of his head.

The worker looks more similar in shape to the queen than the drone, but is much smaller than either. Most of the bees in the colony are workers.

THE DRONE: JUST A GIGOLO

Although the queen mostly lays fertilized eggs, which hatch into female worker bees, she also lays unfertilized eggs, which mature into male drones. These boys are a minority and exist solely to mate with new queens. Day after day they congregate outside the hive and monotonously fly around for hours on end unless the presence of a virgin or newly mated queen snaps them into action. Sadly for the drone, mating is his last act on earth, as the hookup rips his genitals from his body.

An alternative fate is to be booted from the hive and die of starvation or exposure. Ever thrifty, worker bees do not hesitate to banish their brothers from the homestead in lean times when they want fewer mouths to feed. Apart from such untimely deaths, drones live four to eight weeks.

drone

Drones are lovers, not fighters, and they are unable to sting.

THE WORKER: JANE Q. PUBLIC

While the queen toils away laying eggs and the drones are busy trying to breed, the workers do everything else. Depending on their age and the needs of their colony, young worker bees cycle through the following jobs:

» **Nurses** feed the larvae and newborn bees.

» **Attendants** follow the queen, feeding, grooming, and protecting her while helping spread her pheromones around the hive, the scent of which keeps the bees on task.

» **Undertakers** remove dead larvae and pupae from their cells and carry dead adults outside the hive.

» **House bees** remove refuse, including any flotsam and jetsam that get tracked into the hive; use propolis to seal small spaces in the hive and cover any debris that's too large to carry out, like a dead mouse; prep vacant brood cells for a new round of eggs by cleaning out pupal poop and cocoons and polishing the insides of the cells with propolis; unload nectar from foragers, storing it and curing it into honey; make wax to build new comb or to seal comb that is full of honey or brood; ventilate and control the temperature of the hive.

» **Guard bees** live by the sword and die by the sword. On the alert at the hive entrance, they are the first to attack and

These nurse bees tend to the larvae.

sting intruders — such as skunks, bears, and beekeepers — dying in the process. They also patrol for trespassers such as yellow jackets, bumblebees, and honey bees that don't pass the sniff test because they belong to another colony.

» **Foragers** are mature bees a few weeks old that fly outside the hive to scout for and collect nectar, pollen, water, and propolis. They live in the fast lane, flying up to sixty miles a day for several weeks until their wings are literally tattered from the effort and they die of exhaustion.

Bees that hang together in festoons are secreting wax. When you remove a frame, sometimes you will see a string of them as shown below.

tattered wing

The everyday heroine above has logged enough air miles as a forager to wear out her wings.

pollen

pollen

Before bees finally graduate to foraging, they take short practice flights outside the hive to get familiar with their surroundings. You often see them hovering near the hive entrance, whereas seasoned foragers will fly in and out of the hive with a clear sense of purpose. This bee has just returned with a full load of pollen packed onto her legs.

After a bee drops a pollen pellet into a cell, she moistens it with honey and saliva, and then packs it down with her head. Pollen is an important food for brood and the youngest bees, which is why you usually find pollen cells clustered near brood cells.

Home Sweet Home

QUEENS EMIT A PHEROMONE, called *queen substance,* to let the colony know she is on duty. Workers have an alarm pheromone that signals danger. The sweetest scent of all, though, is the Nasonov pheromone, which has a lemony quality and is literally the smell of home. Though all beehives share this scent in general, each one is just different enough for the bees to know their own hive from another.

Workers fan pheromones from their abdomens as a homing beacon for scouts and foragers. The Nasonov gland is here.

The Life Cycle of a Bee

Before bees begin a life of hard work, they have a lot of growing up to do. The road from being merely a twinkle in the queen's eye to becoming a busy bee has three stops.

EGG

The queen drops a single egg into a cell. The egg stands up straight in the cell on the first day, lists like the Tower of Pisa on the second, and topples to its side on the third, after which it hatches into a larva. (Occasionally a new queen will lay more than one egg in a cell as she gets the hang of it, but check for signs of laying workers. See opposite page.)

LARVA

Larvae look like shiny white crescents and get fat so quickly that they molt every day. Workers feed them abundantly for six days, then cap the cells with wax so the larvae can pupate. To make brood cappings, bees recycle some of the wax they pick up from the edges of cells, which accounts for their slightly brown or orange color.

PUPA

Having been fed enough to live on fat reserves, larvae spin a cocoon and pupate, metamorphosing into newborn bees that chew away their cell cappings and join the colony.

An egg is the shape of a kernel of rice but so small it can be hard to see. Larvae are tiny white crescents at first and develop quickly into big, fat grubs.

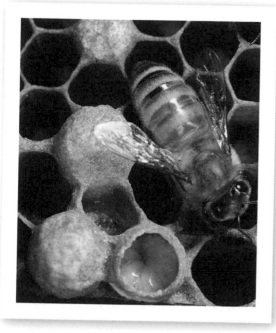

Drones are larger than workers and need more space to develop. The workers literally raise the roofs for drone pupae, giving their brood cells dome-shaped cappings. These cells are usually clustered at the edges of the frame.

Queens normally lay one egg in the bottom of the cell. If there are many eggs in a cell, or the eggs are stuck to the sides of the cell, or eggs are in pollen cells — all of which you see above — it means workers are laying eggs. This can happen when a hive is queenless. Since eggs from workers aren't fertile, they will all develop into drones.

The Ultimate Superfood

ONE OF THE QUEEN'S POWERS is to determine the gender of the bee, by fertilizing an egg with sperm to make a worker — or not, to make a drone. The workers, in turn, have the awe-inspiring responsibility of deciding whether fertilized eggs grow up into more workers, like themselves, or new queens.

The particular food that workers give to developing larvae determines their destiny. At first, all larvae receive a rich, lavish diet of milky fluid, called *royal jelly*, secreted from bees' glands. Larvae that develop into queens continue to enjoy an abundance of this super-food, while larvae that grow into workers switch to different rations.

INCUBATION RATE

Color Key Egg Larva Pupa

QUEEN - → 15½ days

3 days	5½ days	7 days

WORKER - → 21 days

3 days	6 days	12 days

DRONE - → 24 days

3 days	6½ days	14½ days

Bee Diary

We cut this drone pupa out of a cell. It has the form of a bee but is still white and squishy like a larva.

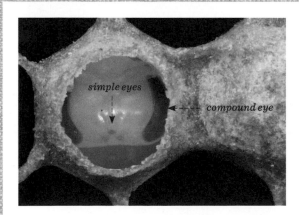

simple eyes

compound eye

This pupa is further along in its development, and the eyes are turning black. Note the two large compound eyes at the sides of the head and the three simple eyes in the center. As the pupa develops, the whole body will turn black.

The Life Cycle of a Colony

The reproduction of bees sustains life within the hive from day to day, but a colony is a superorganism — one whole made of many interdependent individuals. So how does this superorganism reproduce, and what happens when the queen dies?

SWARMS

When a strong, healthy colony matures, population density can become vexing: traffic jams, cramped living quarters, too much competition for resources . . . you may know the feeling. An escape to the quiet of the country begins to sound like a really good idea, and crowded bees plan their exit strategy by raising new queens even while the current queen still reigns.

The bees rear as many as forty queens at once. The first to arrive on the scene seeks out the others, destroying queen larvae and pupae and stinging any other competitors until only she, one virgin queen, remains.

The original queen, meanwhile, has stopped or slowed egg laying and is eating less food, slimming down so she will be able to fly again. When the new queens are close to hatching, the old queen leaves with half the bees and sets up a new colony elsewhere.

Everybody wins but the beekeeper. Swarming is how colonies reproduce in the wild, but if the swarm flies out of your hive, you've just lost half your bees. Giving bees more space so they don't swarm or controlling the split of the hive is a key skill of the beekeeper's trade. (See chapter 7, page 99.)

SUPERSEDURES

Neither age nor illness is kind to queens. If she can't lay as many eggs as she used to, or the level of her pheromones drops, she'll soon be staring down a *coup d'état* (properly known as supersedure) as her bees raise a host of challengers to replace her.

If a queen dies without warning, the colony must have eggs and three-day-old or younger larvae to feed royal jelly and raise into queens.

Bees typically have what it takes to survive this emergency, but presumably, they get a bit panicky. Beekeepers have identified a distinctive roaring sound in queenless hives: the orphans' collective wail. *Queenright* is a term used to describe a colony that has a laying queen.

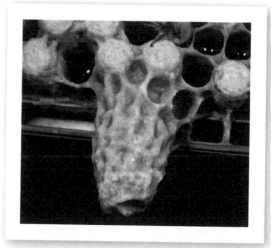

Queen cells cultivated for a swarm usually hang from the bottom of the frame, while those cultivated for supersedure are usually on the face of the comb. All queen cells hang vertically and have a distinctive peanut-shell length and surface texture.

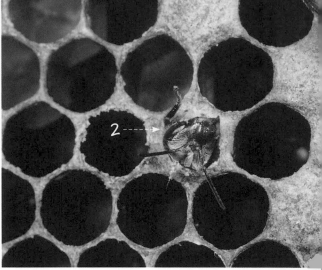

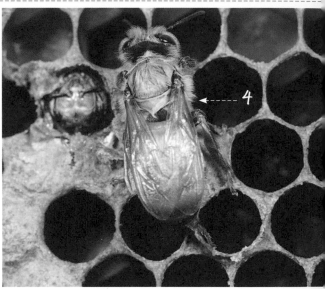

Bee Diary

As we grew in confidence as beekeepers, we became able to observe the hive patiently. One of our favorite moments is seeing the birth of a bee.

1. Here she chews through the capping.
2. Her head emerges.
3. She pulls herself out.
4. Newborns look downier than mature adults, with soft light-colored hairs on their upper bodies.

chapter 4

How Do I Get Started?

How Many Hives?

Your most basic choice is how many colonies to set up. Some people feel to the marrow of their bones that bees are their life's calling, and they want to begin in a big way. For everyone else I recommend starting with two hives for the following reasons:

» As a new beekeeper you won't know what normal is; with two hives you'll at least know what's different between them. That point of comparison can give you an edge in detecting and solving problems early.

» You can use the resources of a strong hive to save a weak one by moving frames of brood or honey from one to the other as needed.

» With two hives instead of one, you'll have a higher likelihood of success your first year. Beekeeping is a bumpy road, and losing colonies to disease, predation, and operator error is not uncommon.

If you are on a very tight budget, or for any other reason two hives are more than you can handle, starting with one is a perfectly good choice. Unlike chickens or goats, more is not merrier when it comes to the number of colonies in your beeyard. In fact, too many hives in one area increases competition for forage and diminishes the strength and productivity of each colony.

Bee Diary

Getting started is always the hard part for me. What if I make the wrong choices? Where do I even begin? If you relate, then do your research. Read. Ask other beekeepers in your area for guidance. Nothing beats a face-to-face conversation with a talented bee wrangler, and my local beekeeper's association has been a solid-gold mine of information.

That said, don't get hamstrung by too much information when you find that every beekeeper has a slightly different perspective and advice. Have the confidence to plunge right in and make your own choices.

What Kind of Starter Colony?

Buy bees from quality suppliers, ideally from local sources that have bees adapted to local conditions. To find a bee supplier, ask for recommendations from beekeepers in your area, either through a local bee association or through the online forum at Beesource.com. Resist using an Internet search engine, which won't filter for good reputation. Ask the supplier what the policy is for replacing a queen that is missing or dead on arrival. A key choice is whether to start with a package, a nucleus hive (also known as a *nuc*), or a swarm.

PACKAGES

A package is a box that contains one mated queen and a bunch of bees that have been collected from an assortment of different hives. A common size for the package, which is sold by weight, is 3 pounds, which roughly equals ten thousand bees. Since a package does not contain brood, pollen, honey, or drawn comb, the colony will be slow to build in strength and productivity.

People order package bees over the winter and are given an assigned day in the spring for pickup. The lines to collect the bees can be very long!

Our very first package of bees on our very first day as beekeepers.

NUCS

A nuc, short for nucleus hive, is a fully functional starter hive that is ready to install in your hive body. It consists of a laying queen and a small quantity of bees on three to five frames of brood, honey, and pollen. (For definitions of frames and hive bodies, see Starting Equipment on page 58.)

Your supplier's equipment must match what you plan to use. If you want to use medium-sized boxes for your hive bodies, a top-bar hive, or small-cell frames, for example, and the only nucs available to you are standard-cell deeps, then a nuc isn't a good choice.

These are temporary boxes we used to transport our nuc frames from our supplier's beeyard. There are ventilation holes and caps to stop up the entrances while the frames are in transit.

One disadvantage of a nuc is that you are inheriting frames with drawn comb and brood, which may contain diseases and a higher level of mites than you'll find in a package.

Nucs are harder to find than packages. Joining a local bee association will put you in contact with veteran beekeepers in your area, some of whom may have nucs for sale in the spring.

On the other hand, nucs also have significant advantages. With a laying queen, developing brood, and available food stores, your colony will rapidly gain strength and productivity. (See page 92 for more on nucs.)

PACKAGE PROS	NUC PROS	SWARM PROS
Can be introduced to any equipment, such as medium boxes, deep boxes, or top-bar hives	Colony gains strength more quickly and reliably and is more likely to have surplus honey at the end of the first season	Supports genetic diversity in honey bee populations if the colony is feral
Less expensive and more widely available	Less chance of losing disoriented bees during installation	No cost
No brood diseases	If the stock is locally bred and overwintered, bees will be adapted to local weather and forage	Comes from local stock healthy enough to reproduce on a colony level

SWARMS

Though it is less common for beginners to start with swarms, many proponents of natural beekeeping swear by them as the best option. By definition, a swarm is the product of a colony that has vigor enough to reproduce, increasing the chances that the bees have a healthy constitution. Also, only a small number of commercial bee suppliers provide 99 percent of the queens sold in the United States, so your choice of feral bees will support genetic diversity. And then there is the obvious benefit that a swarm is free.

Some bee associations, especially in western regions, have community outreach programs through which they advertise a free service for removing unwanted swarms from private property, thereby saving the bees from extermination. The association then finds new or established beekeepers willing to adopt the colony.

When Should I Install My Bees?

YOU TYPICALLY INSTALL A NEW COLONY of bees in the spring; the exact timing will depend on your area. If you are using a local supplier, they can advise you. If you are ordering your bees by mail, you need to connect with local beekeepers, any of whom will have the answer for you.

Preorder your bees in the winter, preferably as early as December or January. Many suppliers sell out of bees well before the spring season, and after all your anticipation and preparation, you don't want that kind of disappointment.

Matt and Jill Reed

PORTLAND, OREGON

One day a cold and weary bee flew into Matt's kitchen and changed his life forever. He didn't swat and kill his curious guest. He offered her hospitality instead, warming her on a plate, feeding her a few drops of honey, and releasing her outside. She must have made it home, because soon there were more bees knocking on his screen door. Eager to know why this would happen, he began researching honey bees in depth. Bees' ability to communicate with one another about the precise location of a food source is one of the amazing facts he learned.

Since that day a few years ago, Matt and his wife, Jill, have made beekeeping their business. Their company, Bee Thinking LLC, manages more than twenty-five colonies in the Portland metropolitan area and sells top-bar and Warré hives of their own design.

Matt and Jill advocate a hands-off, natural approach to beekeeping, letting the bees build their own natural comb, forgoing treatment or medication of any kind, and feeding colonies as a last resort.

The key to their success? Populating their hives with swarms rather than packages. In their experience, captured swarms build up rapidly and have fewer issues that compromise the health of the colony.

What Race of Bees?

Though there is more than one species of honey bee, *Apis mellifera* is the only one cultivated for pollination and honey production. Within that species there are no fewer than two dozen recognized races with unique characteristics. All domesticated races of bees have been bred for qualities that are convenient to the beekeeper, and at the top of the list is gentleness. Gentle bees are less defensive and less likely to sting.

Worldwide, the most popular race is the Italian honey bee, but I started with Carniolans and I know beekeepers who love their Russians. Other beekeepers raise their own queens trying to breed a local stock that is uniquely suited to their region.

Here's the scoop on some pros and cons, but, as always, talk to a local supplier or other local beekeepers about what they recommend for your area.

ITALIANS

GOOD REP	BAD REP
Colonies consistently raise lots of bees, which make lots of honey.	Strong brood rearing even during autumn can result in overpopulation and colony starvation in regions without winter forage.
Bees are gentle and easy to handle.	Italians have a higher tendency to rob honey from weaker hives.
Bees have only a moderate tendency to swarm.	Bees are easily confused about which hive is their home and may drift to other hives in your yard. Painting hives different colors can help Italians orient themselves.

RUSSIANS

GOOD REP	BAD REP
Russians are dainty eaters and thus overwinter on relatively small honey and pollen stores.	Colonies are subject to swarming, which requires different management strategies.
Bees are mite resistant.	Colonies are slow to build population in the spring.

continued

CARNIOLANS

GOOD REP	BAD REP
Colonies strictly limit population growth in the fall so there will be fewer mouths to feed during the cold season, making them especially well suited to survive in regions with long winters.	Colonies have a higher tendency to swarm during the spring baby boom. Carefully timed maintenance is required to give the bees enough room to suppress their swarming instinct.
Bees have first-rate gentleness.	Carniolans are sensitive to drought and other nectar and pollen shortages, making population numbers less reliably strong.

Even bees within the same race can look quite different. Both of these are Carniolans, but the one on the left is mostly black, while the one on the right has much more gold.

Africanized Honey Bees

AFRICAN HONEY BEES are one race of *Apis mellifera* that is not domesticated. They were accidentally released in Brazil and have interbred with other races, resulting in the catastrophe known as the Africanized honey bee, which plagues parts of South America, Central America, and southern regions of the United States. Africanized honey bees exhibit several behaviors that distinguish them from the gentler European varieties:

» *Intensive brood rearing.* Africanized bees devote a higher percentage of cells to raising brood.

» *Frequent swarming.* They are also very successful at reproducing on a colony level, which accounts for their rapid spread.

» *Usurpation.* Swarms of Africanized bees sometimes move into existing hives of European honey bees, killing the queen and taking over.

» *Defensiveness.* Bees surge out of an open hive and defend their nest aggressively, pursuing intruders over long distances.

Chantal Forster

ALBUQUERQUE, NEW MEXICO

Chantal organizes ABQ Beeks, an association of beekeepers in New Mexico's largest city. There are now about 250 beekeepers in Albuquerque, and they tend to be unusually progressive. Chantal estimates that 80 to 90 percent of them use integrated pest management and 50 percent use top-bar hives, many designed by local beekeeper T. J. Carr. Rather than purchasing bees from large commercial suppliers in Georgia or elsewhere, many local beeks capture swarms or acquire nucs from local provider Zia Queenbee Co., which breeds stock that is genetically selected to suit the southwestern climate.

Like most urban beekeepers, Chantal and her fellow beeks have the advantage of an active and tight-knit community to offer support, in addition to having landscapes with diverse and well-irrigated plantings. One challenge her community faces is that they are within the range of Africanized bees (see box on opposite page), which come up the Rio Grande from Mexico every summer. The beeks check their bees regularly for docility. If they notice unusually aggressive behavior, the Extension office at New Mexico State University will arrange to have the bees genetically tested for Africanized genes and will help dispose of the hive to protect neighbors and the community.

Where Should I Put My Hives?

The question of where to site your hives has a different answer for every yard or home, but here are some basic guidelines:

» **Set the hives on a surface that is level** side to side. A slight tilt forward is okay and can even help with water runoff.

» **Choose a warm, dry location.** Areas that are prone to flooding or habitually damp and cool, such as the bottom of a hill or slope, are not suitable. Spots with too much shade or cold winds are also poor choices. Southern exposure at the hive entrance is ideal.

» **Year-round accessibility** to the hives is important.

» **Be sure there are nectar and pollen sources** for your bees to feed from within a two-mile radius of your location. This is unlikely to be a problem as bee forage is available just about everywhere, even in densely developed urban areas. Access to fresh water is also essential.

» **Give yourself some elbow room.** You can space a pair of hives as close as six inches from each other, but you need several feet of space around one or two sides in order to move equipment around as you're working the hives.

» **Know what predators** (man or beast) are in your area and what protection your bees will need.

Set within a fully enclosed cage at a U-pick apple orchard, this hive is protected from prying fingers and paws.

» **The bees' flight path** will extend in a straight line out of the hive entrance, so don't point the hives toward the neighbor's yard, street traffic, your back door, or your doghouse unless you have a hedge or fence to force the bees high into the air.

In addition to these basics, the following are important considerations. Don't be discouraged, though, if you can't put your hives in a location that's 100 percent ideal. Do what's best when it's possible, and be okay with compromises when it's not.

KEEP IT EASY

During peak times of year, you may need to work your hives often, and when you harvest your honey, you'll be lugging extremely heavy boxes. The best scenario is to have your hives outside your home or a short distance away, so the responsibility of caring for your bees doesn't become burdensome or easy to ignore. Like a garden, it's nice to have it near enough that you can visit it often and know what normal looks like.

KEEP IT SECURE

When honey bees have the opportunity to choose their own home, they pick a spot about 10 feet off the ground. Their instinct to build their home in a high place serves them well. A skunk can be a terrible nuisance at a hive and will lurk outside the entrance to catch bees as they come out, rolling each one on the ground before popping the bee into its mouth like a bonbon. Humans, predators of everything under the sun, frequently vandalize hives out of ignorance or malice. Rooftops provide safety from such larger pests. A building, house, or garage with a flat roof is an excellent location for your hives and one reason beekeeping is such a good fit in urban areas.

Even if you aren't a rooftop beekeeper, raising your hives at least 18 inches off the ground will keep your bees out of the reach of most skunks.

While keeping your hives well off the ground is an effective strategy for deterring predators, the height makes it harder to reach boxes full of honey.

Bears are a weightier problem, and if you are keeping bees in bear country, it is critical to set up an electric fence before you hive your bees. Bears will devastate a hive, ripping it to pieces to eat the bee brood and honey inside. The time and money you invest in beekeeping will vanish in an instant if one of these marauders visits your beeyard, and a fence will be much more effective before the bear has tasted the sweet rewards inside the hive.

An electric shock will deter a bear that is simply sniffing out possibilities but not one that knows from experience that a cornucopia of pleasure is just one quick zap away. For this reason it's important to start with an electric fence rather than resort to it as a remedial measure.

I DON'T KNOW A POSITIVE FROM A NEGATIVE charge, so the prospect of bear-proofing my hives with an electric fence paralyzed me. Fortunately, my husband Mars figured it out and offers this tutorial.

The shopping list for a 12′ × 8′ enclosure:

- Eight 5-foot steel T-posts (fiberglass step-in posts are easier to install but aren't sturdy enough for a permanent fence)
- Screw-on, plastic T-post insulators
- 14-gauge aluminum wire
- 6- or 12-volt marine battery and fencer, or a solar unit (automotive batteries do not provide the constant load of electricity that is necessary)
- Solid copper ground rod
- Gates
- Tester

Prep and Layout

» *Allow at least 3 feet of space* between the fence and your beeyard equipment, providing enough distance that the bear can't reach through and grab the hive boxes or the fence battery. A 12′ × 8′ enclosure should be sufficient for two hives.

» *Use 5-foot steel T-posts* as your fence posts and plastic insulators to keep the wire from touching the metal posts. Space the posts no more than 6 feet apart so the wire doesn't sag. We used eight T-posts for our 12′ × 8′ enclosure: one in each corner and one on each side.

» *Clear grass, weeds, leaves, and any debris* in an 18-inch perimeter around the fence. This best practice guarantees you've created a circuit between the ground and the positive charge, should the bear touch even a single wire. You definitely don't want anything from the ground, such as a twig, touching your wire, as it will short out your battery.

Setup

» *Feed a strand of 14-gauge aluminum wire* through each plastic insulator on your posts. Space the insulators 8 inches apart, starting with one 6 inches from the ground, which is low enough to keep a skunk from walking underneath the fence. You should have five to six strands of wire creating the fence walls. Adding X patterns from the top corner of each post to the bottom of the next adds strength and makes it harder for a sneaky bear to slip a paw through without touching a wire.

Remember to turn the fence back on when you exit. It's easy to forget!

» **Control and monitor the charge** of your fence with a weather-sealed fencer and power it with a marine battery, which is sealed and does not require maintenance. Raise your battery and fencer off the ground to keep the battery from discharging, and cover them as protection against the elements. We keep them on top of the box that our first package of bees arrived in and cover them with a plastic tub. Solar units are a good option for beeyards that aren't at home, so you don't need to worry about recharging the battery.

» **Ground the fence** with a 4-foot-long solid copper rod connected to the negative terminal of your fencer, and drive the rod 2 or 3 feet into the ground. Wrap a wire around the positive terminal and run it to the second lowest strand on the fence itself. Any lower and a tall weed can easily short the fence out. *Note*: If you live in a dry climate, you can't rely on the moisture in the ground to conduct electricity to the ground rod, so you will need to wire alternating positive and negative strands.

» **Periodically test your voltage** with a tester that you hook onto one of the wire strands. A probe at the other end goes into the ground and it should read 5,000 to 6,000 volts. If it is lower than that, either you need to charge your battery or you have a short somewhere in the system.

» **Make a gate** so you can get into your enclosure. First create a loop at the end of each wire and secure it to the insulator so it doesn't pull through. At the other end of the wire, attach an electric fence gate, which looks like a handle with a hook. The hooks of the gates lock onto the loops of wire you created. You will need one gate for each strand of your fence.

Living with Your Fence

Turn off the fence when you are working your hives, so you don't bump into it and get a shock. Exercise caution by educating your own children or neighborhood children about the dangers of the fence. Although it won't deliver a shock big enough to seriously harm a person or pet, it does hurt, which is really the whole point.

A couple of years before we got bees, a bear visited our bird feeder. An electric fence around our hives will keep her away from the bees if she returns.

Steven Cameron
SAN FRANCISCO, CALIFORNIA

Steven Cameron has kept bees since 1977. Currently, he has three hives in San Francisco on land managed by Friends of Alemany Farms. Nestled between the freeway and subsidized housing projects, the property is a community garden, its produce sold to economically disadvantaged Bayview/Hunter's Point residents for $1 per bag.

For Steven, the biggest challenge of keeping bees in the city is vandalism and theft. Kids throw rocks at the hives, trying to knock them over. He even found honey supers beginning to disappear, so he now keeps all his hives chained and bolted to their platform.

With the luxury of a mild climate, Steven's bees are active year-round, except when it's raining. The only time he occasionally needs to feed them is November, in between the bloom times of San Francisco's two widespread varieties of eucalyptus trees.

Within the hilly terrain of San Francisco, it can be hard to find level ground. Steven keeps his hives steady on a platform built into a hillside slope. Vandals and thieves have forced him to chain his hives to the stand.

KEEP THE JONESES HAPPY

Although honey bees might lay down their lives in an act of heroic resistance when you open up the hive, they rarely sting people unless they are defending their home. Playing a game of soccer or having a barbecue party in your yard is not likely to distract the bees from their intensely focused work. Even so, you don't want bees bumping into you, your friends, or, worse still, neighbors and passersby. Once again, the rooftop provides an excellent location, as the bees' flight path will start and end at a height that won't interfere with human traffic.

Another option is to face the entrance of the hives toward a high fence, wall, or dense shrubs, which will force the bees to ascend and descend vertically as they come and go and will generally keep their flight path above head level. If you live within a stone's throw of another home and don't have something like that in your yard, consider making plantings or erecting a barrier to serve that purpose. Be sure the barrier isn't so close that it keeps the hive entrance too dark and cool. Warmth and sun gets the bees out of the hive and into the field for foraging.

- -

Unless your next-door neighbors are excited to welcome bees into the 'hood, respect their anxieties and concerns and keep your hives as far from their property line as possible. The same goes for locating your hives away from public lines of traffic.

KEEP IT COZY . . .

Honey bees have the remarkable ability to control the temperature in the hive. When it is too cold, they cozy up to each other in a close mass and can generate a surprising amount of heat. Even so, like most of us, they prefer to make life as easy as possible, so it's best to locate their home in a spot with southern exposure and a mixture of sun and shade. Your climate, of course, factors into the balance of shade and sun that is ideal for your hives. Generally, lots of full sun is ideal to keep the hives warm and dry, with some shade in the hottest hours of the afternoon. In areas with very hot summers, however, a shadier spot will be better.

Wind chill is another environmental factor to consider. If you've located your hives in a gusty area, erect a barrier on the side of the prevailing winds (see page 125).

. . . AND COOL

Bees not only generate fuel-free heat, but they also have AC. When it's too warm they fan their wings and evaporate drops of water to cool down the hive. For that reason they need continuous access to fresh water during the summer, which can be one of the challenges of rooftop beekeeping. I live across the street from a brook, so that's one thing I can strike off my chore list.

If you don't live near a pond or a stream, there are a number of good options for supplying your bees with water, and the best ones give them a place to land without the danger of drowning. Following are some ideas that can work in a rural, suburban, or urban environment:

» Set a pail full of water near your hives and float bits of wood or cork in it as landing rafts for the bees.

» Let an outdoor faucet drip slowly.

» Buy a feeder from a beekeeping supplier. Feeders that are sold for sugar syrup can just as well be used for plain water.

To ensure that your bees don't wander over to a neighborhood swimming pool for their water supply, you can train them to drink from your water source by sweetening it at first with a small amount of sugar syrup. They will be drawn to the sweetness, and then will always return there even once you've switched to plain water.

KEEP IT TIDY

If you're siting your hives on the ground, clear the area of any grass or weeds that could obstruct the hive entrances. The irony is that people spend inordinate amounts of time and money trying to encourage grass to grow on their lawn, but if you want to get rid of the grass, it's an intractable weed. Herbicides may seem quick and easy, but I don't recommend dumping poison in your beeyard. As a backyard beekeeper with just a couple of hives, I haven't found it very difficult or time-intensive to rake away new growth a few times a year.

This bee sucks rainwater off our chicken coop.

To slow or eliminate weeds that are choking your hives, there are lots of options for covering the ground:

» Paving stones
» Cement pad
» Landscape fabric
» Mulch
» Aluminum flashing
» Wooden pallet

You can also keep your hives on a back porch or a spacious balcony. If your hives are on stands that are high enough, the weed problem may be minimal or even nonexistent.

KEEP IT TOGETHER

Especially with new packages of bees, a phenomenon called *drifting* is common. Drifting happens when bees that belong in one hive migrate to another, sometimes by accident. To help bees find their way home when you have multiple hives close together, you can paint the box fronts different colors or with different patterns. You can also orient the entrances in different directions.

Some drift occurs no matter what you do, but if it happens in dramatic numbers, you may have a poor-quality queen. See chapter 6.

Italian bees are particularly challenged in distinguishing one hive from another based on its placement. Differences in color and pattern will help them out.

KEEP IT FUN

Don't get overwhelmed by all these considerations. The most important things are to cover the basics and to apply common sense and respect in relation to close neighbors so you don't wear out your bees' welcome. Do as much as you can for now, but know that you can fine-tune your setup at any time.

Bee Diary

Our first year of beekeeping, we started with two packages. Each had the same number of bees to begin with, but one soon outstripped the other in strength and numbers, and this is not uncommon. Drift rarely occurs equally in both directions, and you may end up with one hive that's noticeably weaker than the other.

chapter 5

THE GEAR

Starting Equipment

Most beekeepers in industrialized nations use Langstroth hives as described on page 28. This hive style comes in two sizes, the most common of which accommodates ten frames. I use eight-frame hives because the smaller size makes them lighter, but I'm so scrawny that my nickname in my old NYC neighborhood was Olive Oyl. We can't all be Popeye, so I don't sweat it. Suppliers manufacture Langstroth hive boxes in both wood and polystyrene. I use wooden hives out of a preference for natural materials.

The chart on the opposite page shows how heavy boxes filled with honey will be. Think conservatively and honestly about what size you can lift without killing your back.

Buyer beware. To save money you might find some used supplies from another beekeeper who is upgrading or wants out altogether. Be cautious when purchasing used frames or boxes — they could contain hidden diseases. You'll have to weigh that risk against the obvious cost savings.

OUTER COVER: THE ROOF

Choose an outer cover that telescopes over the sides of the top box and is made of wood covered with metal, features that will make the cover last a long time and keep the wet weather out of the hive.

You need a heavy weight, such as a large rock, to place on top of the outer cover to prevent a particularly strong gust of wind from blowing it off and exposing your bees to inclement weather.

Prepping the Hive

WHETHER YOU CHOOSE WOOD or polystyrene hives, treat all the exterior surfaces so your equipment doesn't decay. There is more than one way to skin a cat, and I trust that you'll figure out what works for you, but painting the hives with latex is the most common solution. Leave all interior surfaces unpainted, especially your boxes.

When I picked a color to paint our first boxes, I didn't know that the color red appears black to bees. We like to say that we're running stealth hives.

This overturned bucket protects our fencer.

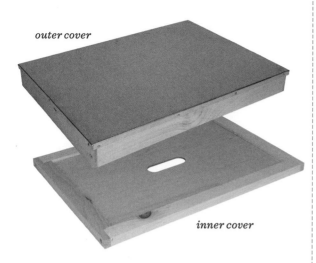

outer cover

inner cover

INNER COVER: THE CEILING

The inner cover is a thin board with a capsule-shaped hole in the center that provides ventilation. Because bees don't recognize anything above the inner cover as being part of their livable house, you can set a feeder on top of it without too much risk of their taking up residence (see pages 74 for feeding and 138 for cleaning supers). You will also blow smoke through the cover's center hole to calm the bees before you fully open up the hive.

The best choice is an inner cover with another hole notched into the front side to provide additional ventilation as well as a secondary entrance and exit for the bees. If your inner cover doesn't have this feature, it is easy to cut a three-quarter-inch notch out of the rim yourself. If the inner cover you get is flat on one side and has a rim on the other, you place the cover rim-side up during the bees' active season and rim-side down during the winter to create air space and cut down on condensation (see page 124).

SUPERS: THE ATTIC

These boxes are storage spaces that your bees will pack full of the surplus honey they hoard. You will relieve your bees of their excess wealth by removing these supers from the hive and collecting the honey for your own enjoyment and nourishment.

Supers come in medium and shallow depths. The advantages of the medium boxes over the shallow are that you can store more honey in the same amount of equipment, and with fewer frames you'll spend less time extracting the honey. If you are likely to injure yourself lifting that weight, though, it's sensible to opt for the shallow boxes.

The number of supers you need will vary depending on how strong your hive is and how abundant the nectar flows are that year. Buy two supers for each hive as a starting point. You're not likely to need more than that in your first year, but if your bees are going gang-busters, you can always extract the honey and put the supers back on for a refill.

HOW STRONG ARE YOU?

APPROXIMATE WEIGHTS OF WOODEN SUPERS FILLED WITH HONEY	
10-frame medium super	55 lb
8-frame medium super	44 lb
10-frame shallow super	40 lb
8-frame shallow super	32 lb

HIVE BODIES: HEARTH AND HOME

The hive bodies are the main living space for your bees. They raise their brood in these boxes and also store honey and pollen here for their own food. Since a fully populated hive would be too heavy for mere mortals to lift, hive bodies are manufactured at more manageable sizes, and you stack them in multiples. Most people use two deeps per hive.

You may occasionally hear hive bodies referred to as deep supers, which can be confusing, as "super" refers to the box that stores the honey you will take. You rarely remove honey from the "deeps," as the bees need a certain amount to sustain themselves.

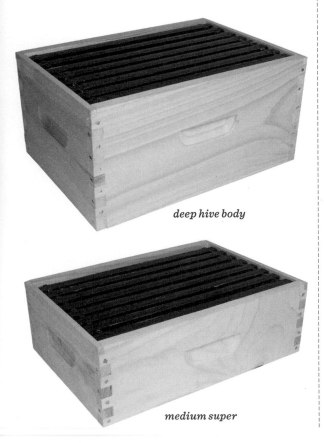

deep hive body

medium super

FRAMES AND FOUNDATION: NO-WASTE SYSTEM

Frames are rectangles made of wood or plastic, providing a structure onto which your bees will draw out wax in the much-celebrated interlocking pattern of hexagons known as *comb*. Each hexagon is called a *cell*, and your most fervent hope as a beekeeper will be that your bees fill as many cells as their busy little bodies can with honey, pollen, and brood.

Each frame should have side bars, ensuring you have the optimal amount of space between them when they are pushed together and touching. Usually, each frame also has a sheet of foundation set within the outer rim. Foundation comes in several different designs, all with the purpose of supporting the wax comb so it's easier for the beekeeper to handle and manage.

Many proponents of natural beekeeping try to avoid bossing the bees around and so advocate using frames without any foundation, letting the colony organize the nest and build the comb more like the way it would in the wild. The theory is the bees instinctively act in their best interest, and overmanagement compromises their ability to manage their well-being. If you go that route, buy wooden frames with wedge-style top bars. Remove the wedge on each frame, rotate the strip of wood ninety degrees, and using nails or staples, reattach it to the top bar to serve as a guide for the comb.

Whenever you work the hive, be sure to leave the frames correctly spaced (see Bee Space, page 28).

Natural Cell Size

IN THE WILD, bees build cells for worker brood that measure 4.4 mm to 5.1 mm. However, most available foundation is made with a cell size of 5.4 mm. These larger cells produce larger bees, but the question is whether bigger is always better. Many beekeepers are going back to the basics of natural cell size by using small-cell foundation or no foundation and report colonies that are better able to resist the pestilential Varroa mite (see page 110). If you choose this route, you need to buy bees from a supplier raising colonies in small-cell hives or take special steps to regress the typical, larger bees commerically sold today. The resources section on page 151 will direct you to books with in-depth information on this topic.

One-piece plastic frames have built-in foundation molded with hexagonal cell patterns (detail shown, bottom right). Such frames are very sturdy and easy to use, though bees get a slower start building comb on the plastic foundation. If you choose plastic, be sure the supplier has coated it with wax.

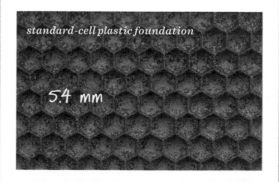

small-cell wax foundation

4.9 mm

standard-cell plastic foundation

5.4 mm

screened bottom board

solid bottom board

wooden entrance reducer

BOTTOM BOARD: THE FLOOR

The bottom board has a rim around the edges that will lift your boxes up a bit to provide the bees' main entrance.

Beekeepers are battling the omnipresent hive pest known as the Varroa mite (see page 110). One tool at your disposal is the Varroa screen: mites fall through the screen to a debris tray below and are unable to reenter. (Clean the tray regularly to discourage wax moth larvae from feeding on the debris.) The screen also provides much-needed ventilation. You can use a Varroa screen by itself as your hive floor or in combination with a solid bottom board.

ENTRANCE REDUCER: THE GATE

In a peaceable world we wouldn't need locks for our doors, but life isn't any safer for bees than it is for us. For starters, when you first install your bees, they will be weak in number and organizational efficiency. Other bees or wasps may take advantage of this weakness

and waltz right into the hive to steal some food. As a beekeeper, it is your responsibility to protect them from robbers, and an entrance reducer will do the trick nicely.

When you buy your bottom board, you may receive a wooden entrance reducer for free. It slots into the space between the Varroa screen and the hive body, blocking most of the entrance and leaving just a small space open. The smaller entrance makes it much easier for the bees to defend their home.

A wooden entrance reducer with a notched hole will be sufficient for deterring robbing insects, but it will not keep mice out once the weather turns cold and they are seeking a cozy burrow. To protect your bees from these interlopers, you will need a metal reducer that is mounted to your equipment. Beekeeping suppliers will have these, or you can make one yourself with some wire mesh or hardware cloth. Half-inch mesh will have holes large enough for a bee to pass through but small enough to lock out pesky rodents.

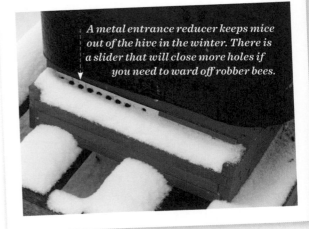

A metal entrance reducer keeps mice out of the hive in the winter. There is a slider that will close more holes if you need to ward off robber bees.

HIVE STAND: THE CELLAR

You need a base to keep your hive off the ground so it stays drier and has an insulating layer of airspace.

Some prefer a base that lifts the boxes high enough that you don't have to bend as much during inspections of the hive bodies. A higher stand also keeps the hive entrance above unmanaged weeds and grass. Others prefer low stands because if there is a bumper crop and the hive gets very tall, the top supers become impossible to reach without a ladder and the hive is more likely to tip in the wind.

You can purchase hive stands or make your own platform. A wooden pallet or cement chimney block also works well.

SMOKER: YOUR BEST FRIEND

The smoker is your best friend. Never work a hive without it! I am by nature an impatient person, and I find the delay of firing up my smoker to be a nuisance. I have tried on more than one occasion to check on my bees without one, and

my last words on the subject are these: don't do it. Smoke calms bees down, and quiet bees will not sting you. See All About Smokers on page 80 for instructions on how to use your smoker.

Smoker Fuel

burlap   pine shavings   pine needles   baling twine

Don't forget that you will need fuel for your smoker as well. There are many options for free or for sale. Your fuel must be dry, free of chemicals, and relatively slow burning. You can use paper to get your fire started, but it burns too hot and fast to be of much use as your primary fuel. Here are some good options:

» Pine shavings, which I happen to have around anyway as bedding for my chickens
» Untreated baling twine
» Untreated burlap
» Pine needles you've foraged from the woods
» Old cotton shirts or dryer lint

I've already stressed the words "chemical free" and "untreated," but let me be explicit: you must be sure any materials you burn will not release noxious fumes into the hive. You merely want to settle the bees down, not bury them six feet under.

Pick up a fireplace lighter from the grocery store, and keep it with your smoker. I kept running back into the house to get a lighter from the kitchen until I finally realized that for a few dollars I could spare myself the trouble.

HIVE TOOL: YOUR OTHER BEST FRIEND

Well, maybe the hive tool is your best friend, not the smoker. It's a tough call. This is your all-purpose tool for working the hive, and you'll want it to stick to you like glue. It pries open your boxes, helps lift out your frames, and scrapes away unwanted propolis and stray comb.

Bee Diary

My husband, Mars, who took all the photographs for this book, couldn't focus his camera while wearing a bee helmet with a rigid veil, so he took a couple of hits to the face before he found an army hat with some soft, flexible mosquito netting built in.

Ouch!

HELMET AND VEIL: BASIC PROTECTION

If you've seen beekeepers working hives without a helmet and veil, don't be tempted to do so yourself. The overwhelming majority of beekeepers, even serious and experienced veterans, consider it irresponsible to work without this minimal amount of protection. Bees go for the head first, and while no sting is fun, a sting to the face is about as fun as a sharp stick in the eye.

GLOVES: COMFORT FOR BEGINNERS

Many, if not most, experienced beekeepers usually work their hives without gloves because you are definitely less clumsy when your hands are bare. I could have put gloves in the section on optional equipment (page 67), but I think for beginning beekeepers they really are essential.

I didn't start this beekeeping adventure with any fear of bees or stings, and in the beginning I worked my hives without gloves. As my colonies gained strength, however, and the bees grew in number, one day I opened up a box and couldn't bring myself to plunge my unprotected hands into a roiling mass of stinging insects. I put gloves on and was happy to have them. Even if you are braver than I, every beekeeper should own a pair of gloves for unusually sticky situations involving particularly annoyed bees.

Beekeeping gloves are made of thick leather or a plastic-coated canvas, and they are very long, extending well above your elbow and cinched at the end of the cuffs with elastic.

Gorilla Suits and Other Inappropriate Dress

PICTURE IN YOUR HEAD a beekeeper and you probably see a guy in a head-to-toe white bodysuit. Fancy outfits like that are a luxury rather than a necessity, and you can save yourself some cash by wearing your street clothes. Not just any street clothes, though. There are definitely dos and don'ts about what to wear to minimize the risk of stings.

Let there be light. Bees are instinctively programmed to attack natural predators such as skunks, bears, and raccoons or anything that resembles those animals. While working the hive, wearing dark or, worse yet, fuzzy clothing, such as wool sweaters, is inadvisable. White clothing is ideal but anything light-colored is good.

Mind the gap. As you work the hive, a lot of bees may get knocked to the ground. To keep them from flying or crawling up your pant legs, tuck your pants into boots or socks if they aren't already fitted at the ankle. You can also cinch your pant legs with elastic. Anyone who wears a skirt to the hive wins a Darwin award. For the same reason you want shirt cuffs that fit closely at the wrist. Tuck your shirt into your pants or cinch it at the waist with the strings of your helmet and veil.

Hang loose. Most heavy clothing, including denim, does not offer any advantage. A stinger is almost a quarter inch long, so you would need to be wearing a thick jacket and ski pants to be sting proof. Such a getup would be murder in the heat of summer. What is more practical is loose clothing. I sometimes find stingers embedded in the white cotton shirt I like to wear to the hive because it provides airspace between the cloth and my skin.

Ignorance is bliss. My bee mentor Tony invited me over to his yard to practice installing a package before my own bees arrived. The day was unusually hot, so I showed up in shorts and a tank top. Tony sized me up and seemed a bit nervous, but with his quiet graciousness, he didn't send me home.

Optional Equipment

The equipment mentioned so far is the minimum you need to get started. You can be a successful beekeeper with just those items. There is a dizzying array of other supplies that you can buy, and much of it will sound really useful and tempting. My advice is to go easy on the wallet. You don't even know yet for sure if you like beekeeping. That said, I am including a short list of additional equipment to consider if your budget will allow nonessentials.

FEEDER

Your bees sometimes need you to help them boost their food supplies, most especially when a package colony is brand-new. As someone who lives in a cold climate where spring arrives late, winter arrives early, and the pickings can be slim in between, I've found the feeder necessary. At the very least, most people use a feeder when installing a package of bees on new foundation. Having a bounty of food motivates the bees to draw out comb, which the queen needs to lay eggs and strengthen the colony's numbers. (See pages 74 on feeding new colonies and 121–122 on supplementing winter stores.)

A feeder is a container that you fill with a specially formulated sugar syrup you make yourself (see page 74), which the bees can eat and store in reserve. I recommend one that you enclose within the hive. A feeder set outside the hive is a welcome mat for robbers who, besides stealing the syrup, may try to enter the hive and further annoy your colony. Your feeder also must be designed so the bees don't drown in the syrup.

You can get or make a bottle or pail feeder that sits above your inner cover, but you will need an additional, empty hive body to enclose it, and as a beginning beekeeper you probably won't have an extra deep kicking around. You can get a division board feeder, which is the size and shape of a frame and takes the place of a frame in your hive body. The drawback to this option is that you have to open up the hive to feed the bees. What I use is a hive-top feeder that is the size and shape of a super and sits above the inner cover so you don't need to disturb your bees to add more syrup.

QUEEN EXCLUDER

A queen excluder is a screen that is slotted with spaces large enough for worker bees to fit through but too small for the queen to pass. You place it between your topmost hive body and your bottommost honey super, ensuring that the queen never lays any eggs in the frames that you'll be extracting honey from.

The use of queen excluders is a polarizing topic. Some beekeepers derisively call them honey excluders because sometimes even worker bees decide they don't want to pass through the screen. Others disagree, not wanting to risk bee brood in the honey frames. They feel it is unhygienic and find it inconvenient at honey harvest to deal with straggling nurse bees that refuse to leave their posts in the supers.

There is no definitive answer, but as a first-year beekeeper, you need to know that bees will not draw out comb above a queen excluder. If by some chance you inherited frames with comb, use the excluder or not as you prefer. Most likely, though, your frames have only foundation, and you will at least need to let the bees draw out the comb before you consider placing an excluder.

If you choose to use an excluder once the comb is drawn out, be very sure the queen is in the bottom boxes. Otherwise, you'll exclude her in the honey super and get the opposite of what you wanted!

Also be sure not to trap drones above the excluder. An inner cover with a notched opening in the rim can serve as an escape route for the big boys. Position your outer cover so it doesn't abut and block the notch.

This grid of white plastic is a queen excluder.

SLATTED RACK

This is a very short box fitted with a series of wooden slats, and it sits between your Varroa screen and your bottommost hive body. As well as providing a layer of ventilation and insulation, it gives the bees a sort of covered porch in which to hang out and makes them feel they have a bit of elbow room, which is useful in preventing swarms (see page 37). The slats line up with the frames, allowing a clear path for mites and debris to fall through to the Varroa screen.

BEE SUIT

You probably have clothes in your closet that are suitable for wearing to the hive (see page 66). Coveralls or jackets providing protection from stings are not at all necessary. After my first season of beekeeping, I put a combination jacket and screened hood on my Christmas list, because I had found the basic helmet and veil somewhat cumbersome. Every time I leaned over at a deep angle, the helmet would slip off my head.

VARROA MITE CONTROL

The pervasive and unavoidable parasite called the Varroa mite is the bane of most beekeepers' existence. Varroa can kill your bees, even in the first year, if you don't manage the level of infestation. I discuss practices and treatments for mite control in chapter 7 (pages 109–113), and when you decide what approach you'll take, you can add any products or equipment you'll need to your shopping list.

HONEY HARVESTING EQUIPMENT

In chapter 8, The First Harvest, we discuss the specific tools and equipment you use to take your honey crop. To spread out your initial outlay of cash, you can wait to purchase those until you know you have some honey to take, which might not even be in the first year. You also might be able to borrow this equipment from your bee mentor, or you can go low-tech with little to no equipment at all. However you manage it, don't be caught at harvest time unprepared to reap the fruit of all your hard work. Plan in advance!

Bee Diary

Sometimes I don't use queen excluders. If I accidentally end up with capped brood in some of my honey frames, it is easy to remove the wax cappings from the honey cells only and leave the brood cells sealed so that baby bee meat never touches my crop.

honey cells

brood cells

Back to the Future: Top-Bar Hives

People kept bees and harvested honey for thousands of years before Rev. L. L. Langstroth invented the modern commercial hive. One ancient style is the top-bar hive, which has enough distinct advantages that some contemporary beekeepers advocate its use as a better alternative.

PROS

» Beekeeping with top-bar hives costs very little and has never gone out of favor in nations where people have scant resources.

» Working a top-bar hive requires no heavy lifting, since you're dealing with one very long box rather than narrow boxes stacked on top of each other.

» You can work on small sections of the hive at a time with far less disruption to the bees. The increasing popularity of top-bar hives in the United States is largely due to the fact that the less you disturb the bees, the less stress you put on the colony, which in turn increases its health and resistance to disease.

» Someone with moderate building skills and a few tools can build a DIY top-bar hive.

CONS

» The comb is not built within frames, so trying to use an extractor to remove the honey is experimental. Most top-bar beekeepers crush the honey out, which is more time-consuming, produces less, and precludes reusing the comb.

» A top-bar box can't be extended to encourage bees to hoard ever more honey in the way that you can keep adding supers on top of a Langstroth hive, so the potential yield is diminished.

» There are still relatively few experienced beekeepers with personal knowledge of top-bar hives. Since mentorship is so important to a beginning beekeeper, it is harder to use methods and equipment that possibly no one else you know can advise you on.

A precept of industrialized food production is that maximizing yields and minimizing labor is more important than nurturing the well-being of the animals we raise or sustaining the health of the earth as a whole. As more of us question the hierarchy of these values and look back to the wisdom and farming methods of previous generations, momentum is building to experiment more with top-bar beekeeping.

Some top-bar hives have sloped sides.

chapter 6

THE FIRST MONTH

You Did It!

You ordered your bees, right? If not, what are you waiting for? (See chapter 4.) Now we're getting to the fun stuff — your up-close-and-personal experience with live bees. Study this chapter before your first day as a bee-keeper. Nothing can completely prepare you for the first time you handle bees, but you'll be more ready and steady if you've done your homework.

ARE YOU PREPARED?

You may be the relaxed, casual type — and that's cool — but here's a heads-up: starting with bees requires a smidge of advance planning. Before your bees come home with you, at the very least you need to accomplish the following:

Paint your hive components, because you can't do it once the bees are in there! (See page 58.)

Select a good site for your hives. It's possible to move them later, but it's enough of a nuisance that you want to avoid it. (See page 48.)

Acquire your beekeeping gear. Don't count on going shopping at a supplier's warehouse when you pick up your bees. Hundreds of people will be showing up at the same time and not only will the supplier be crowded and busy, but a lot of items may also be out of stock. (See chapter 5.)

THE DAY BEFORE

If you haven't already, stack up your hive components in the following order, from bottom to top:
- ❯ Hive stand
- ❯ Solid bottom board (optional)
- ❯ Debris tray
- ❯ Varroa screen

Bee Diary

I started going to meetings at my local bee association before I ever got bees of my own. I met my bee mentor Tony there, and he invited me and some other newbies to come to his house to watch him install some new packages. I had read about how to introduce bees to a new hive, but the text came to life and seemed so much easier once I saw it in action. Even better than watching a demo, Tony let me install one of the bee packages myself! I learn best by doing, rather than listening or reading, so this was a huge help. With that hands-on experience under my belt, I was much more relaxed and confident the following month, when my own packages arrived and I introduced them to the hives.

- Slatted rack (optional)
- Hive body filled with frames
- Inner cover
- Outer cover

Have on hand the following tools and supplies:
- Veil
- Gloves
- Smoker and fuel
- Hive tool
- Pliers (for removing the feeding can from the package)
- Nail (for making a starter hole in the queen cage)
- Entrance reducer
- Feeder, hive-top or other
- Sugar syrup for feeder (see page 74)
- Spray bottle filled with syrup (optional)

The wide boards on the slatted rack and Varroa screen face front.

The hum of ten thousand bees in a package is thrilling. Their trademark buzz comes from the rapid beating of their wings. Bees have no voices.

Feeding 101

BEES DON'T NEED US. They raise their own young, feed themselves, clean house, and are generally self-sufficient. However, there are circumstances in which a colony is weak or starving, and if you take a laissez-faire approach, you could lose the time and money you've invested in it.

Spring Feeding

When starting a package of bees on new frames, you need to supply sugar syrup in a feeder. Forage may be slim in the spring, and easy access to ample food helps to strengthen a weak colony. Consumption of sugar syrup will stimulate young bees to make wax and draw out comb, which is the building block of life in the colony. Without comb, the queen can't lay eggs and the workers can't store food that they forage themselves.

To make early-season sugar syrup, mix water with regular table sugar in a 1:1 ratio. Run the tap as hot as possible before measuring out the water. If the sugar doesn't completely dissolve when you add the water, heat the mixture on the stove without letting it boil.

Keep filling the feeder over the following weeks, or even months, until you see capped stores and there are flowers blooming in the landscape. Try not to let the feeder run dry. You are aiming for steady abundance, not a roller-coaster cycle of boom and bust.

Fall Feeding

If you live in a climate where forage isn't available year-round, you need to watch your colony's food stores to be sure they have enough to last them through the lean times. To boost fall stores, feed the bees in a 2:1 ratio of sugar to water. The lower water content will help them cure the syrup to the moisture level of honey. If too much water remains when the weather turns really cold, the syrup will freeze and be of no use.

Tips for Success

> Never use brown sugar, raw sugar, corn syrup, or anything other than pure cane sugar to make syrup.

> Never feed bees store-bought honey as it may harbor disease.

> Add a teaspoon of apple cider vinegar to one gallon of syrup to prevent mold from growing.

> Replace any moldy or fermented syrup.

> If you heat the syrup on the stove, don't let it boil. Boiling may invert the sugar and cause dysentery in your bees.

> If using plastic frames, spritz them with sugar syrup just before introducing the bees to the hive. The syrup will encourage the workers to draw out comb.

> One pound of sugar fills two cups. Knowing this makes it easy to measure your water without measuring your sugar first.

DAY 1

Ready, Set, Action!

Package bees are seriously stressed out. Besides having been shaken out of their home into a strange box with an unfamiliar queen, they have been bounced and banged around, often over long distances. Be kind to them, and get them into a hive as soon as possible. If the day that you pick them up is cold or rainy, keep the box in a cool, dark, dry place while you wait for better weather.

HIVING THE BEES

1. Assemble your tools, light your smoker (see page 81), and don your veil. You can try working without gloves, but have them near, in case you decide you want them. You probably won't need to calm your bees with smoke on day one but lighting it is good practice.

2. Remove the outer and inner covers and feeder from your assembled hive. Also remove the four center frames from your hive body.

3. Fill the feeder with sugar syrup, and set aside. *Tip:* If you lightly spritz some sugar syrup on the bees through the metal grate of the package box, the ones that get misted will be less likely to fly when you pour them into the hive body.

4. Set the package close to your open hive. Using your hive tool, pry the lid off the top of the box, which will reveal the queen cage and feeder can. The bees have been eating sugar syrup from this can as they journeyed to their new home. Remove the queen cage. Some bees will escape from the box. Keep calm and carry on.

5. Inspect the queen cage, and make sure the queen is alive. If she is not alive (this is rare), contact your supplier immediately and schedule a hasty replacement.

The queen is the largest bee in the cage.

Without Foundation

IF YOU HAVE DECIDED to use foundationless frames, stop at step 11. Place the cage on the Varroa screen before gently reinserting the frames in the hive body; then skip to step 14. This variation is important because a queen cage hanging in between foundationless frames is an obstruction muddling the bees' ability to build comb in straight, parallel sheets.

Beekeepers working with foundationless frames or top-bar hives must be especially vigilant about monitoring comb production and quickly removing any comb that isn't straight and centered. If one sheet of comb is askew, the next will follow the same untidy curve, and so on, ruining your ability to lift out each sheet singly.

6. Using pliers, remove the feeder can. Quickly set the wooden lid back on the top of the box to cover the now-exposed holes. *Tip:* Fewer bees will escape if you give the box a sharp thump on the ground before removing the syrup can.

8. Wiggle a nail through the candy plug, being extremely careful not to impale or stab the queen. The bees will eat through the candy to release the queen, and this hole gives them a head start. Set the queen cage aside for a moment. If it's cool outside, put her in a shirt pocket to keep her warm.

7. Peel back (but don't remove) the metal disk with your pliers to reveal a hole in the queen cage that is plugged with sugar candy. Keep one side of the disk attached to the cage.

9. Pick up the package box, and give it one solid slam on the ground to knock the bees to the bottom. Don't be timid. You want them to lose their footing for the critical next step.

10. In one steady motion, remove or let the lid fall off, turn the box over, and pour the bees into the center well of your hive body. Shake and tilt the box to dump out as many as you can.

12. Gently reinsert the center frames in the hive body. Work slowly and carefully so you don't crush the bees. The frames will settle as the bees move.

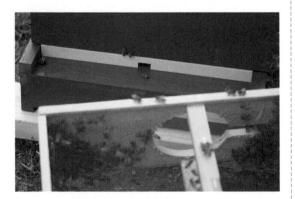

11. Prop the package box so the opening faces the hive entrance and the remaining bees can easily find their way in. The lemony pheromone smell of the others will lure them home.

As you pour the bees into the hive, some will fly out and some will stay in the package box. This is okay.

entrance reducer

13. Insert the queen cage between two frames that are closest to the center of the box. Rest the metal disk on the tops of the frames, so the cage is hanging in the hive. Be sure worker bees have access to the screened sides of the cage.

14. Replace the inner cover, the feeder filled with syrup, and, finally, the outer cover. Insert the entrance reducer to protect your new colony from robber bees.

This mass of bees is in the center well of the hive after being poured from the package.

All About Smokers

Beekeepers all over the world for thousands of years have used smoke to calm bees. In less industrialized times people used torches (and still do in some places), but a smoker is a relatively modern tool that serves the same purpose and is safer for both you and the bees. It has a canister to contain the burning fuel and a bellows to pump smoke out of the nozzle in the lid.

The beauty of smoke as a management tool is that it's elemental and yet works well on a couple different levels:

» As soon as you crack open the hive, bees will release an alarm pheromone that signals danger, and depending on the conditions and temperament of the colony, the troops may attack and defend. Smoke's strong odor helps to mask these pheromones and dials down the threat alerts.

» Where there is smoke, there is fire, and the bees instinctively know this. They will move away from smoke as they would from hot flames, and you can use their sensible good judgment to herd them within the hive.

While smoke works brilliantly, smokers are actually quite touchy, always seeming to burn out at the most inconvenient time. It takes a bit of skill and practice to manage one so it's ready for action when you need it.

HOW TO LIGHT AND FUEL A SMOKER

Light your smoker *before* you approach your hives. In addition to the smoker, you'll need a lighter or matches; a sheet of newspaper; and dry, chemical-free, slow-burning fuel, such as pine shavings, baling twine, burlap, or dried leaf litter (see page 63). Have plenty of extra fuel close at hand, as it's very possible you'll need to stoke the fire in the middle of your hive inspection.

1. Lightly wad up a half sheet of newspaper. Don't crumple it too tightly. You want air to circulate through the wad. Light the paper, and drop it to the bottom of the smoker's canister. I don't like to handle burning paper, so I put the newspaper into the canister first and use a twelve-inch fireplace lighter.

2. Gently puff the smoker bellows to spread the flames. Vigorous, strong blasts of air may put the fire out. Poke the paper with your hive tool if the fire needs more encouragement.

3. When you have strong, licking flames, add a small amount of fuel. Keep gently puffing the bellows and slowly adding fuel. If you add a lot of fuel all at once, you may smother the fire. *Tip:* If you want to spread the flames, realize that fire will move in reverse direction to your movements. When you tilt the canister left, the flames will tilt right.

4. When your fire is roaring and the smoke is thick even when you work the bellows hard for a minute, close the lid. You should see smoke pouring out of the hole at the top. If the smoke feels hot to the touch or sparks fly out when you puff the bellows, add a small amount of greenery (such as leaves or grass) to the top of the canister.

A full canister will probably smoke for about thirty minutes. Even if you don't need to keep smoking your bees during your hive inspection, continue to puff the bellows every so often. If the smoker sits unused, the fire peters out. If you see the smoke starting to thin, pause to stock the canister with fresh fuel.

The easiest smoker to light is one with leftover fuel in it, so leave some char in the bottom when you're done. You will occasionally have to empty the smoker to clear the ashes out of the bottom.

HOW TO USE A SMOKER

1. Send a couple of generous puffs of smoke into the hive entrance. Guard bees are positioned there, and they will be the first to raise the alarm once you approach or open the hive. Wait half a minute.

2. Remove the outer cover and puff several times into the hole in the inner cover. Wait another half a minute to give time for the smoke to seep through the hive.

3. Remove the inner cover and prop it near, but not blocking, the hive entrance. Smoke the tops of the frames to force the bees down into the hive.

You can begin your hive inspection, continuing to smoke the bees while you work, to move them out of your way. *Note:* Smoke is miraculous if used well, but too much smoke will have the opposite effect and create a confused clamor only time will subdue.

Use a smoker every time you work a hive. Every time. Don't get lazy. Don't be rushed. If you don't have time to light your smoker, you don't have time to inspect the hive.

HOW TO EXTINGUISH A SMOKER

Be cautious with the smoldering remains inside your smoker. When I'm done at the bee yard, I grab a handful of grass off the lawn, twist it tight, and stuff it into the lid's nozzle to choke out the fire. I then turn the hot smoker on its side on the concrete floor of my garage a good distance from anything flammable. Some people use a cork or a tight wad of crumpled aluminum foil.

Another option is to dump the smoker fuel out and drench it with water. Think about where you are dumping it, though. Can you guarantee that grass or leaves, for instance, won't hide a spark that may jump to life after you leave?

Bee Diary

One member of our bee club takes her car to her beeyard. After working with bees for the first time, she didn't know what to do with her hot, lit smoker, so she held it out the window as she drove home!

Another bee club veteran tells the story of leaving an uncorked smoker in his truck while he ran into a restaurant to grab a quick bite. By the time he got back to the parking lot, the fire department was breaking through the passenger-side window! Someone had reported seeing smoke and assumed his truck was on fire.

I don't want to rob you of an anecdote of your own, but smokers are dangerous, so have a plan for how to put out your fire and store your tool safely (see above).

WEEK 1

Watching and Waiting

Congratulations: you are officially a beekeeper! On behalf of all beekeepers, let me welcome you to the club. It's exciting, and I know you want to peek on your bees to see how they are doing, but you need to give them space. The bitter truth is that every time you bust open the hive, it shocks them and sets them back a bit in their progress, so don't be overbearing. One inspection a week at first is the sweet spot.

As you wait patiently for the first opportunity to peek inside, watch the hive entrance. There is a lot to see and learn, even from the outside.

Do you see foragers returning with full pollen baskets? (See page 33.) Since pollen is brood food, you probably have an active queen. Do the abdomens of bees returning to the hive look more distended than those leaving the hive? Then their honey stomachs are full of nectar.

Checking a hive-top feeder doesn't count as an inspection. At first you may need to fill it more than once a week so it doesn't run dry.

Do you see wasps trying to infiltrate the hive to steal hard-earned stores?

Checking for Vital Signs

A week after you've installed your package bees, the queen should be out of her cage and laying eggs. Let's be sure:

1. Grab your hive tool, light your smoker, and don your veil. Depending on your comfort level, do or don't put on your gloves. Approach and work the hive from the side or back. Blocking the hive entrance provokes alarm and disrupts the bees' flight path.

2. Smoke the hive. (See page 82.)

3. Remove the queen cage, and check that it's empty. If there are bees inside, they are probably workers that wandered in. If so, set the cage near the hive entrance and let them wander back out.

4. Remove any burr comb that workers have built around the queen cage or on the tops of the frames.

5. Remove an outside frame and set it aside, propping it against the hive stand. Taking one frame out of the box lets you slide the others around during your inspection. When bees are building a new nest, they work from the center frames to the sides, so at this point there probably won't be much action on the outskirts.

Success! The queen cage is empty.

Use your hive tool to scrape burr comb off the tops of the frames.

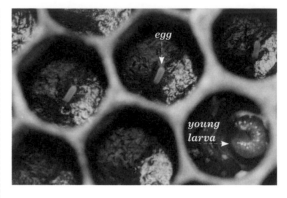

egg

young larva

6. Pull out a single frame, and inspect both sides. Hold the frame over the hive body as you inspect it, in case the queen falls off. It would be so much better for her to fall into the box than onto the ground.

7. Once you have looked at the frame, place it back into the hive in the exact same position and orientation in which you found it. Don't reorder the frames, and don't turn them around. Bees are very methodical planners, and if you change their organization haphazardly, you are doing them a disservice.

8. Repeat steps 6 and 7 until you have found eggs and larvae, a happy sign that the queen is alive and at work.

When you find eggs or larvae, you can breathe a sigh of relief. The queen is present and active.

New wax is pure white. With age and use it will yellow. Comb that has been home to brood will turn a deep orangey brown over time.

TIPS FOR A WELL-ORDERED INSPECTION

There are so many dos and don'ts in beekeeping that it's hard for a beginner to remember them all. Though I usually could remember the importance of keeping my frames in the right order and direction, once I started pulling them out and turning them around to inspect them, I'd easily lose my place. It got even trickier once the colonies grew and each hive had multiple boxes. Here are some lessons learned:

» I set only one frame down outside the box, which creates room to slide the other frames around. I create a wide space around any subsequent frame I'm lifting out for inspection so it's clear where it needs to be put back.

» I mark my frames with a Sharpie pen at one end of the top bar, and I take care that the marks are always lined up. If I get mixed up about which direction to put the frame back in, I have a visual guide.

» I also mark the upper right corner of the fronts of all my boxes with a black dot so I know the direction they need to face when I stack them back up after an inspection.

The Grim Keeper

BEEKEEPERS SEND SOME BEES to meet their Maker every time they work the hive. It's inevitable. With all my heart I hate to put a box or a cover back on and hear the crunch of crushed bees as they get trapped between the parts. Using smoke to clear the tops of the boxes before you stack up the hive will minimize the loss of life.

Another unpleasantness is killing bees by dragging one frame across another when you remove it. (Crushing the queen in this way is called *rolling the queen*.) Before removing a frame, slide the surrounding ones away to make space. Then lift it straight out of the box so it doesn't touch the others. Your hive tool is handy for getting a frame's lug up so you can grab it.

This innocent is trapped and crushed between two hive boxes.

Queen Replacement

You don't see any eggs or larvae at your first inspection? Wait a few more days. If the queen still isn't laying a week after she was released, you have a world of trouble. The situation is serious but not unsolvable. You need a new queen. Contact your supplier to see if they will supply you with a replacement free of charge. There are a number of reasons the queen may be missing, dead, or failing to lay eggs. In any case, your bees need a new one on the double. It will hurt to do it, but pay for the overnight shipping.

STRANGER DANGER

Remember that your bees will not know this new queen. If you release her directly into the hive, they will consider her an enemy and will probably kill her. You need to opt for a slower release.

1. Leave the queen cage corked shut. Do not expose the candy.

2. Prepare a small amount of sugar syrup (page 74), and put it in a spray bottle.

3. Prep yourself and your tools for working the hive, and smoke it.

4. For a sweet introduction, mist the queen cage lightly with sugar syrup.

5. Insert the queen cage between two center frames, screen-side down to direct the queen's pheromones into the hive. Press the cage into some drawn comb to secure it.

To kill an unfamiliar queen, a cluster of bees forms a tight ball around her, making it so hot she suffocates.

Bee Diary

When I had a queen delivered to my office by overnight mail, it caused quite a stir. It's not every day an animal arrives in a box, with air holes punched into the cardboard. My coworkers were curious to have an up-close look at a live queen bee.

STATE OF THE UNION

After five days the queen's pheromones should have spread through the hive, and the bees will be ready to accept her.

1. Prep yourself and your tools, including a spray bottle with sugar syrup, and puff smoke through the entrance and the inner cover.

2. Open the hive, but do not use any more smoke on the frames. Observe how the bees behave near the queen cage.

> Are bees swarming over the cage, grabbing or biting the screen? Try brushing them lightly away. Do they quickly return? These are signs of aggression, so close up the hive and come back in a few more days.

> Are bees passing over the cage casually or solicitously inserting their antennae toward the queen? If so, proceed to step 3.

3. Remove the queen cage, and spray it lightly with sugar syrup. This helps keep the queen from flying away when you release her.

4. Hold the cage close to the center of the hive, and peel the screen back with your hive tool. Make sure the queen crawls down between the frames.

5. Come back in another week and repeat Checking for Vital Signs (page 85).

With the screen open, tip the cage toward the frames and watch the queen crawl into the nest.

WEEK 2

Free Fall

Any colony of bees started from a package, even one that is progressing like clockwork, is in free fall, desperately scrambling to survive its first month. The math is cruel:

» A worker lives for four to six weeks, so right now the youngest ones in your colony are already middle-aged. Some are older, and others have already died.

» From the moment your queen lays her first eggs, it takes twenty-one days (three weeks) for those eggs to hatch into workers.

Add it up. The package bees are dying, with no new bees to replace them for another week at a bare minimum, probably a few days longer. The population is actually decreasing, rather than increasing.

The good news is that bees do not experience fear or discouragement. They are working diligently and steadfastly and will most likely survive this crisis. They are drawing out comb so the queen has somewhere to lay eggs, and if she is worth her salt, she is building her

productivity. She starts off slowly but will eventually lay up to two thousand eggs a day.

The sugar syrup you are feeding the bees is lightening their workload during these troubled times. While sugar from nectar and syrup powers the grown-ups with carbs, the younglings need a lot of protein, which they get from pollen collected by foragers.

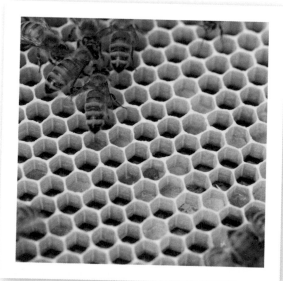

After two weeks, these package bees have begun to draw out comb and fill some cells with brood, honey, and pollen.

The bees are capping the colony's very first cells of honey. Note the holes in the unfinished cappings.----->

This brood has probably just been capped and will hatch in another ten or so days. Note the difference in color and texture between the cappings of brood cells and those of honey cells above.

Nucs 101

ONE ADVANTAGE OF A NUCLEUS colony over a package is that it doesn't experience the same free fall in population. In addition to a starter quantity of workers, a nuc has three to five frames of drawn comb, so the bees aren't building the nest from scratch. There is also a queen that has already been in the colony at least a month, so brood is developing in all stages and new bees are hatching every day. Though a nucleus colony starts off small, it builds more quickly than a package and is more likely to thrive in its first season.

Nucs are harder to find than packages, and their availability often travels by word of mouth. It's best to get a recommendation from a trusted beekeeper who will vouch for the quality of a potential supplier's nucs.

Finding the Right Fit

If you are planning to keep bees with anything other than deep Langstroth hive bodies and standard-cell foundation, you have extra legwork to find a supplier that matches your equipment. If you want to use medium boxes, don't get a nuc with deep frames. Similarly, if you want to raise bees on small cells, find a supplier with small-cell bees. Nucs aren't an option at all for top-bar beekeepers.

Asking Questions

If you find someone who has a nuc available, ask some questions to gauge the quality.

» *How many frames is the nuc?* Three is a minimum, but five is better.

» *How many frames of brood will be included?* Brood should be on at least three frames, with stores of honey and pollen.

» *What is the type of queen?* Good to know, even if you aren't shopping for a particular type, such as Carniolan or Italian.

» *How old is the queen and how long has she been in the nuc?* She should be either newly mated or raised last year.

» *How long has the queen been in the nuc?* She should have been active in the colony for at least a month, not newly introduced in a cage.

» *How will I transport the bees home?* Providers have different arrangements for delivering bees, so find out what to expect. Your supplier may loan you a wooden nuc box that you return once you've installed the frames in your own deep; they may provide you with an inexpensive corrugated cardboard or plastic box; or they may expect you to bring a box and exchange frames on site. (Note that a nuc box has ventilation holes and a sealable entrance, though you can jury-rig a regular hive body to be fully sealed while transporting the bees.)

Set your hive up so it's ready in advance for installing the bees. Bring your smoker, fuel, veil, gloves, and hive tool to your supplier's beeyard. You will need them if you are able to inspect the nucs.

These nucs are waiting for beekeepers to claim them.

Inspecting the Nuc

When you pick up a nuc from the supplier, you may have an opportunity to choose from among a number of available colonies. Look for hives where the bees sit quietly on the frames rather than rushing out of the box in agitation. Open the hive, smoke the bees, and inspect the frames for eggs, larvae, and a good pattern of capped brood. The more brood you see, the better. If you don't see eggs and larvae, you may not have a mated queen.

Installing the Nuc

Set the nuc next to your hive. Bees can easily overheat in the small confines of a nuc box, so open the entrance as soon as you get to your beeyard, and then follow these steps:

1. Remove some frames from the deep hive body — enough to make room for the nuc frames plus one or two more to give you space to move without rolling the bees.

2. Smoke the nuc box entrance, open the nuc, and smoke the bees on the frames. Remove one of the outside frames slowly and gently. Transfer it to the deep.

3. Continue to transfer the frames one by one, keeping them in the same orientation and position as they were in the nuc box.

4. Replace enough frames from the deep to fill the box. Position those frames at the outside edges, so the nuc frames are in the center.

5. Put on the inner cover, the feeder filled with syrup, and the outer cover. Insert an entrance reducer.

You are done! Wait at least a week to check on them, and follow the same management routine as you would for a package colony.

Be Like Fonzie

Look for the continued expansion of comb and the presence of brood as well as pollen and honey stores. Your new colonies are still under stress, so keep the visits brief and not any more frequent than once a week.

Now that you've handled bees at least a couple of times, it's time to summon the Spirit of The Fonz and channel coolness. Practice working the hive calmly, quietly, slowly, and gently. Bees react defensively to quick motions, loud noises, and rough handling. Our third week of keeping bees, my husband Mars was yelling at our son to put his veil back on. A bee flew into Mars's mouth and stung him. Double thumbs down.

The Littlest Beekeepers

OUR SON, XAVIER, has an eager interest in nature and science, and I thought keeping bees would be a great family activity. We acquired our first hives when he was eight years old, and while he really enjoyed learning about bees, we soon learned he wasn't old enough for more than a capsule-sized commitment.

Working the hive is intense, focused work, and you can't do it right and look after young kids at the same time. If they show an interest in your hobby, take them out for brief observations. Making them stay at the hive longer than they want to be there will stress you out. You'll pass that stress along to the bees, and the situation can quickly escalate to Code Red.

Sharing the hobby with motivated older kids, on the other hand, can be a great bonding experience.

Fair Weather Friends

There is light at the end of the tunnel. The dark days of your package bees are at an end, and the immediate future is bright. New bees should be hatching by now, and the colony will begin to gain strength in numbers. Early spring is passing, and more flowers are blooming, so forage is more abundant. Continue to feed the bees if they are taking the sugar syrup. Don't stop until they have capped stores and floral nectar is abundant.

The only advantage of a weak, disorganized colony is that it is less defensive. Since you dress like a pro and calm the bees with smoke, you may even have escaped stings so far, but there's more to know about avoiding defensive sallies.

When it's sunny and warm, much of the colony is outside the hive, foraging for food. Cold, rain, and wind keep them home. Under those conditions not only are there more bees present to defend the colony if you open it up, but they also sense a clear and present danger to the safety of their brood when you let the damp and chill in. For the health of the nest and to avoid defensive stings, work the hive in pleasant weather.

Bees aren't active outside the hive 24/7, even when the weather is nice. Foragers return to the nest by evening, and they don't venture out again until the sun has warmed up the world. Hopefully, you've situated your hives in a spot with maximum hours of sunlight. If it's shady, especially in the morning, bees will take longer to roll out of bed and get to work. Since they aren't morning people, and you don't want to mess with rudely awakened bees, it works to everyone's benefit for you to wait until a decent hour to call on the hive. With some patient observation you can determine what time things really start cooking out there.

Bees are sensitive to odors, so avoid perfumes, colognes, or any other strong scents when you're working the hive.

My glove caught this stinger.

Ouch!

You've dressed appropriately; smoked the bees enough but not too much; you're calm, quiet, and collected and working the hive gently and slowly; it's the middle of the day, and the weather is warm, still, and sunny. You are the picture of perfection, and yet a bee stings you. It happens. There are forces beyond your control. Maybe a predator such as a skunk was annoying the hive last night. Maybe your noisy neighbor's leaf blower upset them. Or maybe you aren't perfect. Many of us aren't.

So you have a sting. What do you do? If you can see the stinger, flick it off. Even with the stinger removed, you now have a bull's-eye of alarm pheromones on your body. Other bees may narrow in on the danger zone their sister has marked, each delivering an additional sting. Blow smoke or rub dirt on the area to conceal the smell. Later on, after you're finished at the hive, you can take an antihistamine if you need to alleviate swelling.

One sting leads to another if you don't conceal the smell of alarm pheromones.

Bee Diary

After my first day working a hive in shorts and a tank top, I learned always to suit up for the job. Almost always. After ignorance comes a little knowledge, and as they say, a little knowledge is a dangerous thing. I was in a hurry one drizzly, cool day, so I hustled over to a hive early in the morning wearing a raincoat over a loose shirt that didn't cover my waist. I wasn't wearing a veil and didn't bring a smoker. See how many mistakes you can count.

My much wiser and more cautious better half tried to stop me, but I told him the heavy vinyl of the raincoat was stingproof. We were both right. No bee could sting through the raincoat, but a number of bees flew inside it from the bottom, got trapped at my exposed waist, and panicked. I felt as if I had been hit with a lead pipe.

chapter 7

THE FIRST SEASON

Expect the Unexpected

I wish I could walk with you through every week that follows and tell you exactly what to expect and do within a precise calendar. For better or worse beekeeping is not predictable enough to make that possible. Many variables play into the development of a colony, including:

» **Region.** Hive management tasks are tied to the region where you keep your bees. The time of year when flowers are producing nectar, for example, is geographically specific.

» **Weather.** Even within one region, varying weather patterns from year to year have a huge influence on your colonies. A particularly cold, wet spring or hot, dry summer, for example, will impact the growth of plants and affect the foraging success of bees.

» **Quality of queen.** Even colonies with the same external conditions will fare better or worse, depending on how productive the queen is and the quality of the bees she produces. Some bees, for example, might be more hardy and disease resistant.

» **Skill of beekeeper.** And, of course, some colonies will fare better or worse depending on how skillful their keeper is. But this isn't an invitation to beat yourself up over weak hives or take all the credit for strong ones. There is a lot that just isn't within your control.

With such unpredictability from colony to colony and year to year, I can't lay down a weekly schedule, but I can set out a seasonal one.

Bee Diary

I've never kept journals, and I'm a terrible note taker, but I recommend that you do as I say, not as I do, and keep records. Assign each hive a name or number, then note the dates you inspect each, as well as the observations you make and actions you take.

My family and I like to name our beehives. We pick city names because each hive is a community. Our first two colonies were named Sparta and Carthage. Sparta got its name because the bees in that colony were always more hawkish than the other. I think most of my stings the first year came from those Spartan soldiers. Then of course we wanted to pair it with another ancient city, and we picked Carthage because it was founded and first ruled by a queen.

Manifest Destiny

You started your colony in a single hive body, and so far you've been inspecting for the presence of the queen and brood. At some point in the spring or summer, you'll notice the box begin to fill up. When two-thirds of the frames are in use, with honey, pollen, or brood stored in the cells, your bees are ready to expand, and it's time to add a second deep box on top of the first. When two-thirds of the second box is in use, add a super on top of the deeps, and so on.

If bees don't have enough space to accommodate their growing population, they will swarm.

ALL IN GOOD TIME
More is not better when it comes to adding boxes. It's not good practice to expand the hive before the colony needs the space. A really big house is hard for a small family to keep clean, so pests such as wax moths can move in and wreak havoc. Additionally, bees gravitate to the center of the hive, and if you stack another box on too soon, they may move up and develop the frames in the center, leaving the outside frames in the box below unused (the chimney effect).

Remember that I advised you not to rearrange the frames? There are exceptions to the rule, one of which is to fix the chimney effect. If your bees are not drawing out comb on the outer frames, you can swap those with frames closer to the center that have just comb or comb and honey but no brood. Do not move frames with brood. Brood frames must remain together where the nurses cluster over and attend to them. Any brood that is separated from the heart of the nest may be neglected or catch a chill and die.

A frame with comb but no brood can be swapped with an undeveloped frame to attract the colony's interest in moving out as well as up.

Honey Flows

Experienced beekeepers are keenly aware of which flowers provide bees with nectar, when those flowers are in bloom, and whether the right balance of rain and shine has encouraged them to bloom in abundance. In beekeeping jargon, when there is a bounty of nectar for bees to forage, a *honey flow* is said to be on.

Honey flows are regional to the highest degree, and the only effective way to figure out if you're having one is local observation. If you've made connections with other beekeepers in your area, you can take advantage of their seasoned surveillance. Otherwise, here are some signs to look for:

» **Lots of blooming nectar sources.** This isn't as easy as it sounds. Not every flower is a nectar source, so you need to educate yourself about which ones are destinations for honey bees. Also, bees can travel several miles in foraging expeditions, so it's not a matter of looking out your window and seeing what's up in the backyard. Keep an eye out as you travel around your town, noticing where honey bees are gathering.

Bee Diary

Our second month of beekeeping, we found a suspicious double cup-shaped structure in one of our hives. We panicked that our bees were making swarm cells, so I brought a photo of it that week to a bee club meeting. An old-timer explained that these "queen cups" are often found on the lower edges of combs. Usually, they are empty (as mine were), but even if larvae are present, it does not always mean the workers will follow through in rearing a queen. A cell that becomes very long, on the other hand, has a developing queen inside, as the workers build space to accommodate her larger body.

- **Lots of nectar in the comb.** This is a great indication of a honey flow, but if you're feeding the bees, you won't know whether you're seeing sugar syrup or floral nectar. If the bees stop taking the syrup, however, you know a whopper of a honey flow is on and they are getting all they need from the land. That's worth celebrating!

- **Lots of fresh, white wax.** For a colony of bees, plentiful nectar and pollen is equivalent to wealth. Flush with resources, they begin a building boom and rapidly expand the comb to make room for more brood and more food.

- **Lots of foraging.** During a honey flow, your hive will bustle with activity. Bees come and go from morning to evening in greater numbers than you can count.

- **Rapid weight gain.** Give your hive a heft test each time you visit it. Heave it up a smidge by the handle of the bottom box. Weight gain is always a good thing, and a rapid increase probably means a honey flow is on.

- **Strong, sweet scent.** If walking by your hive feels like passing through a cloud of ambrosia, there is probably lots of honey ripening inside. Happy day.

A honey flow can last days or weeks, depending on the bloom time of the nectar-rich flowers. If you've positively identified that a honey flow is on, pause or slow down the frequency of your inspections so you don't hamper the colony's burst in productivity. Your main task at this point is to provide enough space for the bees to store their bounty.

Depending on where you live and what the weather is doing, your first honey flow might be in the spring.

WHAT'S IN BLOOM?

By no means a complete list, here are some nectar sources you might find in your area:

- Acacia
- Alfalfa
- Apple
- Aster
- Avocado
- Basswood
- Blackberry
- Black locust
- Blueberry
- Buckwheat
- Chestnut
- Clover
- Cranberry
- Eucalyptus
- Fireweed
- Goldenrod
- Huckleberry
- Lavender
- Linden
- Orange
- Poplar
- Pumpkin
- Raspberry
- Rosemary
- Sage
- Saw palmetto
- Silkweed
- Snowberry
- Sunflower
- Thyme

Honey Dearth

LIKE HONEY FLOWS, times of dearth are tied to the natural growing season, which is highly regional. In some hot, arid climates, nectar-rich flowers are scarce for a couple of summer months; in very mild climates, there is a brief period of winter scarcity. Other variables such as droughts or floods may also decimate local sources of forage, creating a period of dearth.

If nectar is scarce, bees may try to steal the stores of other colonies. Italian bees are particularly prone to stealing, which beekeepers call *robbing*. If robbing is taking place, you will see fighting on the landing board, as the guard bees struggle to defend the entrance.

A strong colony has the force to fend off the siege, but a weak one can be devastated. Protect weak colonies during a dearth by limiting access to the hive in the following ways:

- Seal any upper entrances, such as a rim hole in the inner cover or a ventilation hole in an upper box. Duct tape works.
- Keep the lower entrance reduced in size.
- Use a feeder that is inside the hive, so robbers aren't attracted to an external sugar source. If you only have an entrance feeder, make sure sugar syrup isn't leaking and put a reducer next to the feeder so the entrance is on the far side.

If you have a hive open for inspection and robbers arrive, halt the plunder by quickly battening down the hatches, putting the cover back on, and reducing the entrances. Smoke the other colonies in the yard to calm the frenzy. Don't inspect that hive again until a honey flow is on.

Karl Arcuri

AUSTIN, TEXAS

Drought is a major challenge in Texas, but Karl Arcuri hit the jackpot his first year as a beekeeper. The perfect combination of rain and sun struck the land's sweet spot, yielding a nectar bonanza. His single hive, started from a package, had six to seven supers stacked up and he harvested several times that season!

Not every year is so lucky, unfortunately, and Karl has found urban beekeeping especially advantageous in his regional climate, where wildflowers usually bloom only in April and May before drying up under the scorching sun. In the city, though, people water their plants, sustaining the growing season for a much longer period and providing bees with ample forage from diverse nectar sources. Another advantage is that rural farms usually use a lot of pesticides, while he keeps his bees in a neighborhood that's a magnet for people advocating alternative, chemical-free lifestyles.

I love Karl's sense of humor. He names his queens with pop culture references like Large Marge from Pee-Wee's Big Adventure and Ramona and Knives from the Scott Pilgrim graphic novel series. His wife, Gitanjali, and friend Brenna even embellish the hives with art.

SUMMER

Steady as You Go

As the population of the colony expands, there are new challenges to master. If all has gone according to plan, you have lots more bees than you did when you started. If this outward sign of success causes some inner turmoil, don't feel alone. Working a hive of forty to sixty thousand bees is understandably more daunting than working one with ten thousand. Your skills as a beekeeper have been growing along with the size of your colony, but if the former haven't quite kept pace, don't worry, my friend. Stay steady and you'll get there.

A healthy hive has brood in the center of the frame, surrounded by a ring of pollen cells, with honey stored at the outer edges and corners.

DOTTING YOUR I'S

Every couple of weeks, continue to monitor your growing hive for the presence of all stages of bee life, including eggs, larvae, and capped pupae. Ideally you will see a tight arrangement of cells in use, without a high proportion of empty cells mixed among brood cells. Brood of the same age should be grouped together: capped cells together and larvae together, for example.

CROSSING YOUR T'S

Another key to summer management is monitoring storage space within the hive, adding boxes in time to allow for population expansion and honey hoarding. If you are starting a new colony, it's possible you will not be able to harvest much, if any, honey the first year. Though you may see lots of honey in your deeps, if you take it away, your bees could starve during a nectar dearth. What you are waiting for is any surplus stockpiled in the supers.

If you have a particularly strong colony or a particularly strong year for nectar flows, you may find yourself with a super full of capped honey, in which case pass Go and collect your reward (see chapter 8).

AND OTHER, FINER POINTS

Trial and error will boost your proficiency as a beekeeper. Here are some tips I've picked up along the way.

» If inspecting all layers of the hive, work from the bottom up, so you aren't driving more and more bees into the boxes as you inspect each one.

» When the boxes stick together, use your hive tool to pry them apart. If there is a lot of burr comb holding the frames together, you'll need to twist the boxes apart.

» To help prevent queen loss and so the boxes don't pick up dirt and grass on the ground

Brood of the same age should be grouped together. Here a ring of uncapped larvae surrounds a central circle of capped pupae.

as you remove them during inspection, use the outer cover as a platform. Invert the cover so there are fewer points touching (thus fewer crushed bees). Place the box perpendicular to the cover.

Bee Diary

Mars and I love eccentrics, so we have a lot of affection for the odd folks who cover themselves in bees for kicks or carnival sideshows. Traditionally, you played this stunt by wearing thousands of bees on your face in a cluster resembling a beard. The secret behind the magic was a queen cage tucked under your chin.

Recently, bee bearding has gone to extreme lengths with a Guinness World Records category for "heaviest mantle of bees." Almost the whole body is covered in bees to achieve this record, with the mass attracted by use of a synthetic queen pheromone. Some people even compete to hold as many bees as possible inside the mouth!

There is bee bearding, and then there are bearding bees. Our first summer keeping bees we wondered if the mass of workers grouped on the front of the hive in the evening was a warning sign of a readying swarm. But no, this is called "bearding," and bees do it as a response to hot weather, much like sitting for a spell on the verandah to escape the stuffy house inside.

Baby, It's Hot Outside

I'm lucky that summers aren't very hot where I live (but unlucky that winters are very cold). If the mercury sits above 90°F in your yard for months at a time, give the bees some help to fight the heat.

» **Ventilation.** Air circulation is key to a hive's health in extreme temperatures, whether hot or cold. A screened bottom board (see page 62) or a slatted rack (see page 68) will ventilate the hive from the bottom.

» **Sunscreen.** You can't really slather your hives in zinc oxide, but you can paint them white, which will reflect rather than absorb much of the sun's heat.

» **Water.** Bees need access to water, which they evaporate to cool the hive (see pages 53–54).

Bees beard the outside of a hive when the weather is hot.

Bee Diary

We never studied the anatomy of a honey bee until we found this dead one outside the hive.

A bee sucks nectar from a blossom with a long, tubelike tongue called a proboscis. She stores it in her honey stomach, an internal organ separate from the stomach that digests the food she eats. When she returns to the hive, house bees will suck out the nectar and ripen it into honey.

To collect pollen, a bee scrapes it from a flower with her hind legs, which are covered in tiny hairs. She spits out a bit of nectar so the pollen sticks together, packs it into a pellet, and presses it into a cavity, called a pollen basket, on each hind leg.

Mandibles are for chewing. For example, workers masticate wax they have secreted from a gland in order to form comb.

ABDOMEN

THORAX

HEAD

Wing

top view

Mandibles

Proboscis

Sting

Pollen basket

Red Alert

When colony collapse disorder (CCD) hits a colony, the bees literally disappear. You open the hive, and no one is home. No dead bees. No signs of disease. Just nothingness. And its scale is shocking. Some commercial beekeepers have lost thousands of hives in the blink of an eye. The losses have been dramatic enough to create a crisis in agriculture, with fewer and fewer colonies available to pollinate crops.

The drama and mystery of CCD is juicy fodder for the media. While media attention is certainly deserved, and I wish it were sustained, there are other scourges affecting the viability of honey bees that no one outside the beekeeping community ever hears about. For backyard beekeepers, public enemy number one is the Varroa mite, not colony collapse disorder. If you have a colony of bees, you have Varroa mites, too. They are inescapable.

There are three basic approaches to dealing with Varroa mites. Pick one, try them all, or use them in combination, but don't ignore the elephant in the room. These little buggers are small in size, but they loom large in a colony's life.

Varroa mites gouge into this adult bee and suck her blood.

Varroa Mites 101

LIKE ALL THE BEST VILLAINS, Varroa mites are bloodsuckers. Few bees die directly from parasitic feeding; most die from contracting viruses that enter through the feeding wound or are transmitted through internal contact with fluids from the mite's mouth.

Life Cycle

Bee brood is critical to the life cycle of the Varroa mite. When there is little brood in the hive, the mite population is low, and conversely, when the colony grows in size, so does the population of Varroa.

A female mite crawls into a brood cell containing a larva and hides at the bottom of the cell where nurse bees are less likely to find and remove her. A mite's first choice as a host is a drone, since he has a slightly longer incubation period than a worker, but she will invade a worker's cell when there aren't enough drones available.

After the larva begins pupating and the cell is capped, the mite comes out of hiding and lays one male egg and four to six female eggs. Both mother and children feed on the bee, which can weaken or kill it.

Daughter mites mate with their brother. The females exit the cell when the bee hatches; the male stays behind in the cell and dies. Each female then bores a hole into a passing nurse bee and feeds on her for a few days. When the mite smells a larva of the right age, she hops off the nurse to invade another brood cell and reproduce.

Signs of Infestation

Here are some signs that Varroa is having a significant impact on your colony. Unfortunately, the same or similar symptoms can result from other diseases. The appendix on page 146 summarizes what to look for in other maladies.

- ❧ Spotty pattern of capped brood cells, indicating that lots of pupae are dying and bees are cleaning them out
- ❧ Lots of discolored, dead larvae
- ❧ Lots of brood cell cappings with small holes, indicating dead pupae
- ❧ Bees that are stunted or have shriveled wings

A clear and proactive way to determine the level of mite infestation is to do a count. For the beginning beekeeper, the easiest method is to use a sticky board in combination with a screened bottom board. You can buy sticky boards commercially or make your own by cutting down a piece of poster board and smearing it generously with a greasy product such as Vaseline. If your screened bottom board came with a tray, you can apply your goop to that.

Place a sticky board under the screened bottom board and wait twenty-four hours. When some bees dislodge mites while grooming, the mites will fall through the hive, through the screen, and stick to the board you've inserted.

If there are more than forty mites on the board when you remove it one full day later, the mite population is at a serious level.

When to Treat

Use nonchemical controls such as drone-comb trapping and sugar dusting when your colony's population is strongest, typically from June through September. These techniques only work as preventive measures, so don't wait for a high mite count to start nonchemical controls and be consistent in their application.

If you opt for chemical controls of any kind, treat only after a mite count indicates threateningly high levels. It is unsafe to eat honey exposed to most miticides, so apply chemicals after you have taken your last harvest of the season. You can't wait too long, though. These chemicals are temperature dependent, so you have to apply them well before the weather turns cool. (*Note:* Mite Away Quick Strips — a formic acid product — is an exception when it comes to food safety and is approved for use with honey supers on.)

Brood should be compact, without a checkerboard of empty cells as shown here.

Stunted growth is another disease associated with Varroa invasion.

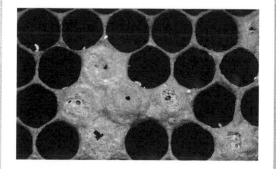

Pupae died in these cells. Undertaker bees have been chewing into the cappings to remove the brood.

This bee contracted deformed wing virus, a symptom of Varroa infestation.

NATURAL SELECTION

Some beekeepers are putting natural selection to work through nonintervention. If you don't treat a colony and it survives the mites for three years or longer, it has qualities that make it stronger than the colony that succumbs the first or second year. Those are the bees you want more of, not the ones you have to prop up artificially.

Your typical bee package from a large commercial supplier will not hold out against mites for long without intervention, but if you seek out people who are breeding survivor stock, you will have a better chance of success with this strategy. Survivor bees have a genetic trait called Varroa-sensitive hygiene, making workers more likely to identify and remove pupae infected with mites. There are also the options of raising Russian honey bees , a mite-resistant strain, or using small-cell or foundationless frames.

MECHANICAL CONTROLS

To help your bees battle Varroa mites, there are chemical-free weapons in the beekeeper's arsenal.

» **Support nutritional health.** You know that when your body is stressed, from poor eating habits, for example, you're much more likely to get sick. Honey and pollen that bees forage from the land are the best food for them. Never take more honey from the hive than the bees need for themselves and assume that you can replace it with sugar syrup. Never feed them honey from the supermarket, which can actually introduce disease.

» **Dust with sugar.** Giving your bees a bath of powdered sugar will dislodge many mites and send them falling through the screened bottom board (see page 115 for details).

» **Trap mites in drone comb.** Mites reproduce most successfully within drone comb and have evolved to seek out the cells of drone brood first, before resorting to worker brood. Use this to your advantage by baiting mites into frames filled with drone pupae that you remove and freeze to kill your quarry (see page 116 for details).

ABCs of IPM

YOU'LL HEAR BEEKEEPERS toss around the phrase *integrated pest management*, or IPM. Here's an abridged definition: a philosophy practiced in agriculture, as well as beekeeping, that seeks to solve the problems of pests and diseases with methods that reduce or eliminate the need for chemicals.

An IPM approach to Varroa mites includes all of the strategies discussed in this book:
» Selecting breeds or genetic traits with a tolerance to mites
» Encouraging healthy populations
» Practicing mechanical controls
» Monitoring mite levels
» Using pesticides as a treatment of last resort

Dusting the hive with powdered sugar

» **Break the brood cycle.** Mites depend on bee brood for their life cycle. If you keep your queen from laying, either by caging her for several weeks or by removing her and letting the bees raise a new queen, you have stopped not only the bees but also the mites from reproducing. This technique is effective for established colonies, not those in their first year.

Mites do not survive without bees as their hosts. When a colony dies from Varroa predation, the mites die also.

CHEMICAL CONTROLS

Chemical control is a touchy subject. People who use hard pesticides often think this is the only efficient and effective way to control Varroa mites. There are considerable downsides, however, including these:

» Because of extensive use, mites have developed resistance to the hard chemicals on the market. Using these chemicals as a preventive measure, rather than to treat a verifiably high level of infestation, is irresponsible and unsustainable.

» Long-term exposure compromises the health of the queen.

» Beeswax absorbs the chemicals, becoming permanently contaminated.

» Miticides in combination with other chemicals (either that you introduce or that the bees bring back to the hive from the landscape) may have unintended effects that compromise a colony's health.

Increasingly popular as an alternative are so-called soft chemicals, including thymol-based essential oils and formic acid pads or strips. Though these products don't kill mites as forcefully as the hard chemicals, they are gentler on the bees, don't pollute the wax, and are thought to be less likely to breed resistance. Nevertheless, no matter what commercially available treatment you use (if any), follow the directions *exactly*. Sloppy practices, such as leaving a treatment on the hive longer than the directions indicate, expose parasites to weakened chemicals that are below a verifiably lethal threshold. Those that survive the exposure are the ones with resistance to that chemical.

Backwards Beekeepers
LOS ANGELES, CALIFORNIA

Laura Stewart, Max Wong, Sue Talbot, and Roberta Kato are friends, neighbors, and fellow associates in Backwards Beekeepers, an organization committed to taking beekeeping back to basics. Renouncing treatments of any kind and most management practices, such as feeding, requeening, and using foundation instead of letting the bees build natural comb, their motto is "Let the bees be bees!" This maxim expresses a strategy of letting colonies alone and the strongest will survive.

A key to the group's success is that they capture native colonies of feral bees, which they've found to be more disease- and pest-resistant than commercial packages. Their bee-rescue hotline offers local residents and businesses a welcome service for capturing swarms that land on their property, and it provides an ample resource of free bees for people who want to start beekeeping or expand their apiaries. It's fun to hear these folks refer to their colonies by where they were trapped, such as Home Depot! Everyone wins, including the bees, which otherwise would probably be eliminated with chemicals.

Sugar and Ice and Everything Nice

A lot of us get into beekeeping because we want to have a positive impact on the environment and an enjoyable intersection with the natural world. Messing with chemicals that can burn your eyes, skin, or lungs if improperly handled probably didn't fit your vision when you started. The good news is that you don't have to use chemicals if you don't want to. There are some Varroa controls as harmless as sugar and ice.

SUGARCOATING THE PROBLEM

Birds are born with an instinct to take dust baths, which loosen mites that have burrowed through their feathers. Most honey bees do not yet have a genetic instinct to groom away Varroa mites, since the parasite has been invading bee colonies only since 1987 and most beekeepers remediate chemically rather than breeding for hygienic traits. Taking the principle of this strategy from other animal species, some beekeepers coat bees with a dusting of powdered sugar. As the bees clean the sugar away, many mites drop off and fall through the screened bottom board. Here's how to do it:

1. At each hive inspection, remove the supers (if you have any on top).

2. Sift confectioner's sugar into the top hive body, making sure that sugar falls in between all of the frames. Use a half cup if there is only one deep or a full cup if there are two deeps.

3. Use a bee brush (see page 129) to sweep sugar off the top bars of the frames.

Sugar dusting removes *only* adult mites that are feeding on adult bees. Two-thirds of the mites are safely nestled inside brood cells and won't be affected by the sugar dusting. A sticky-board mite count taken a week after your first dusting will show the same level of mites that you had before, but this is because new mites have hatched out. What the sticky-board count can't show is that fewer adult females succeeded in moving into cells and laying eggs.

A dusting of confectioner's sugar coats this bee.

PUTTING THE FREEZE ON MITES

Mites prefer to lay eggs in drone cells. Drones take a few days longer than workers to hatch into bees once their cells are capped, which increases the reproductive success of mites threefold. We can use this knowledge to lure the mites into a trap.

The basis of the trap is a frame with drone-comb foundation, which looks much like other plastic foundation except that the cells are larger, signaling the queen to lay drone eggs there. Here's how to do it:

1. Beginning in June or whenever the colony seems strong enough to sacrifice some brood-rearing energy, insert a drone-comb frame close to other brood frames within each deep. To make room you need to remove a regular frame. Choose one that has no or little brood.

2. Wait until the queen lays eggs and many of the cells are capped before removing the frames — three to four weeks. *Timing is critical!* If you don't wait until the cells are capped, you can't guarantee that there are mites inside, but if you wait too long and the mites hatch, you will have reared a bumper crop of parasites, which is worse than doing nothing at all.

3. Put the frames in plastic bags inside your freezer, which will kill the enemy and, sadly, the drones, too.

4. Before the next hive inspection, remove the frames and let them warm to room temperature. Break the cells open with a cappings scratcher and reinsert the frames into the deeps. Worker bees will haul the mortal remains out of the cells and start the cycle all over again. *Tip:* If you have a bird feeder, you can scrape the drone frame clean after freezing and the birds will eat those tasty grubs up!

5. Continue trapping monthly, but stop when brood rearing drops off for the season.

Sugar dusting and drone trapping are most effective when used in combination, since each strategy on its own does not pack enough punch to manage Varroa adequately.

This frame is filled with drone cells.

FALL

Last Harvest

In the United States, even in the southern states, winter-blooming plants are rare, so if you have any harvest at all your first year, the last will be in the fall. (See chapter 8, page 136 for details on harvesting honey.)

My hope is that you'll say no to chemicals, but if you are medicating your hives for mites, nosema, or any other pest or disease, just after the last harvest is the time to do it. Note that medications are often temperature sensitive and require up to six weeks of consistently warm weather to be effective, so don't wait too long for your harvest or the window of opportunity will close. Again, let me stress the vital importance of following a pesticide's directions exactly and using only approved methods.

Bee Diary

Our first year of keeping bees, we freaked out when our hives started smelling like dirty socks. Did they have foulbrood disease? Our bee mentor, Tony, assured us the distinctive scent was ripening goldenrod honey and that it freaked him out his first year, too!

Our first harvest ever was goldenrod honey, and it tasted like apples. Though goldenrod honey is quick to crystallize, and ours turned solid within a week, we enjoyed it in our tea all winter long.

Storing Supers

If bees need to rebuild the comb each year, they produce less for you. When frames are already drawn with comb, the bees can get straight to work filling the cells. Keeping the comb from year to year, then, is like honey in the bank (though you should not reuse comb for more than three years, as it becomes more prone to carrying disease and pesticide loads). After the last harvest you'll have supers with a cache of comb that requires careful storage.

Comb darkens with age and use as bees travel over it with pollen and propolis. Compare the almost-black color of this comb to fresh, white wax.

MICE

Mice sometimes eat and nest in the wax comb in the winter when cozy outdoor burrows with well-stocked larders are scarce. Storing unused frames in sealed spaces is a good idea: stacked hive boxes with an outer cover on top and no entrance at the bottom is one option; an unused refrigerator or chest freezer would work too.

MOLD

Mold grows on comb when the air is warm and humid. The bees will clean a moderate amount of mold off the wax, but it's best to store the frames in a dry location to avoid problems. I use a dehumidifier in my basement and store the frames there.

WAX MOTHS

Often the most significant threat to stored frames, wax moths are a common pest that will destroy the comb (see appendix, page 150). Wax moth eggs may be in your frames even if you've never seen evidence of the insects in your hive. A healthy colony will keep the nuisance in check, but once the super is separated from the bees' diligent attention, it is vulnerable to an infestation. Here are a few different ideas for protecting the comb so you can reuse it next year:

» **Freeze the frames** for at least twenty-four hours to kill any eggs before storing them. Afterward, put the frames into large plastic trash bags and seal with packing tape to prevent later wax moth entry. This is a sure-fire solution that will eliminate further risk.

» **Spray the comb** with *Bacillus thuringiensis* (Bt), a bacterium that attacks the gut of wax moth larvae but is harmless to bees and humans. It is available commercially under the names Thuricide, Dipel, or Certan. Air-dry the frames before storing. Put the frames back in their supers, with sheets of newspaper between the boxes to impede the moths from traveling from one box to the next. This is considered an organic solution.

» **Use paradichlorobenzene** (PDB), sold under the name Para-moth. While these crystals are effective at controlling wax moths, I encourage you to try one of the chemical-free solutions above. In any case, PDB should only be used on frames with empty wax comb. It will ruin pollen or stored honey. After PDB exposure, the wax comb must be aired for several days before reintroducing it to the hive.

» **Keep the frames exposed to light and air** while in storage. Wax moths favor close, dark spaces, so this will minimize the risk.

Keeping your frames exposed to light and air will keep waxworms and moths from eating through the comb.

Stayin' Alive

In the fall the queen begins to cut back dramatically on egg laying, downsizing the colony for the lean months ahead. She also begins rearing "winter bees," which have fatter bodies to sustain themselves through the cold season.

Bees are smart cookies, planning ahead. Take their lead and do the same. If you live in a cold climate, you have to get ready for winter early, since you can't work the hive once the weather is consistently cool. Opening the boxes when the temperature is below 60°F is not ideal, and exposing the brood to temperatures below 50°F is very risky.

You may see workers kicking drones out the door in the fall. They don't want to feed extra mouths through the winter.

FEAST OR FAMINE

Check that your hives have enough stored honey to feed themselves through the winter. How much is enough varies by how long your winter dearth is, but a good rule of thumb for colder climates is to leave the hive with two full deeps. If the size of your colony is robust and there weren't floods, drought, or other adverse environmental conditions, the bees probably collected enough winter stores, but inspect all the frames to be sure, and answer these questions.

Are all the frames chock-full of honey, pollen, or brood? If your answer is yes, you're probably in good shape.

If the colony doesn't have enough stored honey, it will starve during a dearth. This bee reaches for food but the cells are empty.

Have the bees filled the first deep but haven't yet filled the second? If you have extra frames of honey from other colonies, you can put them in, in place of the empties. If the frames are sized for supers rather than deeps, open the honey cells with a cappings scratcher (see page 130) and place them in a super above the inner cover. Since bees usually don't recognize anything above the inner cover as being part of the hive, they will move the honey from these frames into the deeps below. It will seem like manna from heaven.

If you don't have extra honey, feed them with a 2:1 sugar syrup solution (see page 74). Be sure to do this early enough that the bees have time to cure the syrup, reducing the water content to a level where it won't freeze. That's why feeding is a fall activity, not a winter one.

This bee laps up some spilled honey. Note the proboscis inside the honey drop.

proboscis

Did the bees use the first deep earlier in the summer, then move up to the second box and neglect the bottom? Bees have an instinct to move up to find stores, so it's important that the cluster start the winter in the bottom box. You can guarantee this by making sure the brood is concentrated in the center of the bottom deep with honey stores at the sides and in the top deep. If the bees didn't get the memo, make an exception to the rule of not moving the frames around.

Again, if both deeps aren't full, you'll need to feed the bees to supplement their winter stores or they may starve to death.

NOT FIT FOR MAN NOR BEAST
When the nights turn cold, everyone wants to hunker down in a snuggery. Mice get positively desperate and are exceedingly resourceful in wriggling their way into private homes, including beehives. While I sympathize, the bees and I aren't really willing to share, especially since mice help themselves to more than just nest space. As soon as the fall flow is over, it's time to install a sturdy entrance reducer that is screwed securely to the hive furniture. Mice can chew through wood, so metal is a better choice. If you make your own entrance reducer out of hardware cloth, you can leave it on year-round.

Critters start early in securing a cold-weather nest, so don't wait for winter to reduce the entrance.

WINTER

Winter Cluster

When the temperature falls to 57°F, the bees huddle together in a compact mass called a *winter cluster*. If you could see a cross section of the hive, the cluster would look like a sphere, separated into slices by the frames. The cluster concentrates body heat, much like snuggling with your family on a cold night. Bees have a warming exercise, too: by vibrating their wing muscles, which appropriately looks a lot like shivering, the bees increase their body heat. Of course, it's a lot warmer in the center of the cluster than it is on the outside. Always team players, bees rotate positions so no one freezes her butt off for too long.

When it's merely cool, the cluster is loose and large. The colder the temperature, the tighter they huddle, so the sphere gets smaller. If a severe cold spell lasts too long or the cluster is small because of low population numbers, the bees may eat all the honey within their tight mass and not be able to move to other stores. It's a common tragedy for a colony of bees to starve within inches of an abundance of food. Honey, honey, everywhere, nor any drop to drink.

Worker bees that are born in the winter, when less or no forage is available, live for months rather than weeks because they don't wear themselves out with incessant labor.

Huddled Masses Yearning to Breathe Free

Bees do a gangbuster job of keeping themselves warm. All that activity and body heat in the cluster make it downright humid. In fact, what kills a colony more often than cold temperatures outside the hive is lack of ventilation inside. If the humidity isn't vented, moisture will condense and either drip onto the bees, or freeze, making your hive boxes into iceboxes. The following techniques will keep the hive fresh and the bees dry:

» **Use a screened bottom board.** In addition to helping manage Varroa mites, the screen provides ventilation at the lower level.

» **Improve drainage** by inserting a small wedge underneath the bottom board, tilting the hive forward about an inch. Be sure the support is wide enough to keep the hive stable. It shouldn't be tippy. Siting the hives on a slight slope is a permanent solution.

» **Turn your inner cover** so the side with the rim hole is down. If your inner cover doesn't have a ventilation hole in the rim, make one by cutting out a notch. Alternatively, insert a small wedge to prop the inner cover up about half an inch. This hole or space not only vents the hive at the top but also provides the bees with an emergency escape, should the main entrance become blocked by snow or the accumulation of dead bees.

Bee Diary

Bees don't willingly foul their house with feces and can hold it in for long periods of time. A warm winter day is a huge relief, and bees will take bathroom breaks, called "cleansing flights." Sometimes bees come out for a cleansing flight and get so cold they can't make it home.

Protect and Serve

Now is a good time to take a critical look at how you've sited the hive and evaluate if you're serving your bees as well as you can. At the most basic level, they need plenty of sun. When temperatures are cold, the more sunlight the better. Too much shade is never ideal, but in winter it can spell disaster.

WINDBREAKS

If your beeyard is exposed to winter gusts, a windbreak on the north side will help the colony survive. When you were deciding on a site for the hives, you probably considered the following features to minimize wind exposure:

- Evergreen trees or shrubs
- Densely planted deciduous trees or shrubs
- Permanent hardscape, such as fences, walls, or buildings
- Hillsides or other topographical features

If those aren't viable options, you can still make temporary windbreaks by doing one of the following:

- Stacking hay or straw bales
- Assembling a temporary fence from scrap plywood (though temporary, it must be very secure so strong winter winds don't blow it over and take the hive down with it)

Remember to keep a brick or heavy rock on the outer cover so the top of the hive doesn't blow off.

Xavier rescues a bee from freezing. After he gently scooped it up and breathed on it, the bee perked up and flew away.

INSULATION

Sunlight, sufficient honey stores, a sturdy entrance reducer, and ventilation are essential for winter management. Wind protection increases your colony's chance of survival if your winters are particularly cold. A lot of people stop there, but if you want to go the extra mile, you can also insulate the hive. Wrapping the hive in tar paper not only offers a layer of wind protection, but its dark color has the added advantage of absorbing heat. Be sure to leave the entrances clear.

PEACE AND QUIET

Now it's time to let the bees alone. Don't snap open the hive and disturb the cluster unless you have an incredibly compelling reason. Remember that low temperatures chill the brood.

There's no need to clear snow away from the hive if there is a top entrance for the bees to use in cleansing flights. In fact, snow is a great insulator, because of the air spaces between its loosely packed crystals.

If you are lucky enough to have some warm winter days, you can periodically check that the bees aren't running out of stores. As long as the days or nights don't get below freezing, you can feed sugar syrup. If temperatures do get below freezing, you can feed emergency rations of fondant, a heavy icing sugar often used for decorating wedding cakes. Place a sheet of rolled fondant on top of the frames, and add an empty super so the cover will still fit snugly on the box.

If you're anxious to know whether the bees are still alive, give the box a good rap and put your ear up to it. The knock will stir the cluster, and you should hear buzzing.

chapter 8

THE FIRST HARVEST

One Fine Day (Maybe Next Year)

Many new colonies won't have a surplus of honey to share with you the first year. It can be disappointing to delay the gratification of a honey harvest, but keep it in perspective. The bees have worked their tails off building a colony from the ground up. An extra super of honey might not be in the cards. And remember, remove frames of honey only from the supers, not the deeps, or you'll starve the bees.

So this year or next, you will have a crop of capped honey in a super. Hurray for you! If harvesting honey is about the journey rather than the destination for you, by all means take a frame or frames out of the super at any time and reap what the bees have sown. Personally, I find extracting honey messy and time-consuming in a not wholly pleasurable way, so I prefer to wait until the season is at a close and I can do it all at once.

If you don't have any frames of honey to harvest, indulge yourself in a couple of spoonfuls from the deeps. One taste will convince you the wait is worth it.

Harvesting Tools and Equipment

As mentioned, it's important to assemble your harvesting equipment well in advance, or to make firm arrangements to borrow them from a beekeeping friend. Here are your options.

BEE BRUSH

A bee brush has soft bristles that gently sweep the bees off the frames when you want to take the supers away for harvest. Given how inexpensive these brushes are, consider getting one with the rest of your basic equipment, as they can also be useful during routine hive inspections.

BEE ESCAPE OR FUME BOARD

The very cheapest way of removing the bees from your honey frames is with a simple bee brush, but bee escapes and fume boards are inexpensive options as well. (To compare the pros and cons of each, see page 135.)

Bee escapes come in many different designs, all with the same function of letting the bees exit the supers easily while making it difficult for them to reenter.

A fume board is treated with a chemical odor that bees find repellent, driving them out of the super.

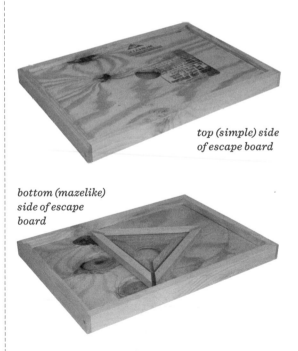

top (simple) side of escape board

bottom (mazelike) side of escape board

If you get a triangle escape board like this one, be sure the simple side of the exit faces the supers and the mazelike side faces the deeps, so you don't trap the bees in the very place you want them to leave.

UNCAPPING TOOL

Once the bees have cured the nectar into honey, they seal or *cap* the cells of honeycomb with wax. To remove the honey you will need to open those cappings. An inexpensive and effective option is a cappings scratcher, also called an uncapping fork.

Another option is an uncapping knife, which has serrated edges and an offset handle to make it easier on your wrist. The deluxe versions are electric and have heated blades that slice through the wax as if it were butter. The economy versions are cold but still do the trick. The cheapest option of all is a bread knife, which you probably already have in your kitchen, but it's going to be a royal pain unless it's very sharp.

With an uncapping fork, you can either scratch the surface of the cappings or slip the tines of the fork under the cappings to lift them away.

UNCAPPING TANK

While you're opening the cells, you'll need a container to collect the honey-soaked wax. An uncapping tank looks like a storage tub you probably already own, but it's different in some key ways. The most important difference is that the tank is made of food-grade plastic, and it has a grate and spigot at the bottom, so you can collect the honey that drains off the cappings. There is also a crossbar at the top where you can rest the frame as you're working.

EXTRACTOR

An extractor holds your frames and spins the honey out once you've removed the cappings. Even cheap, hand-cranked models cost enough to make you pause, and the cheapest motorized models cost more than a beginner should contemplate spending. Consider asking a veteran beekeeper if you can rent or borrow his or her equipment; or forgo mechanical extraction altogether. A great option for the beginner or the beekeeper with frames of foundationless comb is to manually crush and strain the honeycomb.

Kids love to spin the extractor.

MAXANT
HONEY PROCESSING EQUIPMENT

HONEY SIEVE AND BUCKET

As the honey is whipping out of the frames, it drains out of the extractor through a separate sieve and into a honey bucket. The sieve sits on top of the bucket and filters out stray bits of wax. Made of food-grade plastic, the bucket has a spigot, called a *gate*, at the bottom, and it's the last stop for your honey before you bottle it.

HONEY BOTTLES

From wee 8-ounce sampler jars to bulk-sized jugs, there is a wide range of containers you can buy to store and distribute the precious fruits of your labor. If you are harvesting just a few frames your first year, you can really use any jars, bottles, or other containers you already have stockpiled in your kitchen.

You open the gate of the honey bucket to bottle your harvest.

Old-Fashioned Is the New Fashion

THERE'S A LOT TO BE SAID for keeping life simple. Extractors and tanks may sound like a lot of cash and bother. There are other ways!

If you use comb honey foundation or no foundation in your supers, you can cut the honeycomb out with a knife and store the sections in containers. To serve and eat, cut a chunk of comb off and drizzle the honey onto your food or in your tea. Popping a nugget in your mouth is also a delicious candy treat.

Comb honey looks very special and pretty in a jar and makes a great gift, but if you use honey in any quantity, such as in baking, it's not very practical. When you'd prefer to collect your honey separate from the wax, you can still bypass the expense of a mechanical extractor by manually crushing and straining the comb. Manual extraction is possible with any kind of foundation or foundationless frames. See page 137 for details.

The Honey House

Rule number one is to extract the honey indoors. If you try to harvest it outside, the bees will find you and take the honey back, drop by drop. Professional beekeepers have honey houses, dedicated buildings for extracting and handling honey. I know you don't have one, and neither do we, but wherever you extract and handle the honey should be a temporarily dedicated place, whether it's your basement, garage, or kitchen.

You need space to spread out, get messy, and take your time. Depending on your equipment, even a small harvest of one super could take half a day. If you are using your kitchen, understand that the harvest will mean a significant disruption to the daily routine. Keep the peace by talking it through with everyone in the household.

Bee Diary

Although honey is miraculously robust and never spoils, it is sensitive to moisture and readily absorbs water from the air. If exposed to humidity for too long, honey will ferment. My garage is off limits as a honey house because it's muggy in warm weather. We run dehumidifiers in the basement, so it's a better spot in that respect. However, it also means lugging backbreakingly heavy supers down a flight of stairs; there isn't a sink close by for cleanup; and the temperature is cooler than ideal (warm honey extracts and filters more quickly). We extract in our kitchen. You'll have to weigh the pros and cons of your own options.

The Honey Heist

Rule number two is to get the bees out of the super before you extract the honey. Imagine the chaos of bees flying at you from the frames while you're trying to harvest. Here are the most common methods for backyard beekeepers.

Bee brush. One at a time, give each frame a solid knock on the ground in front of the hive to dislodge most of the bees. Gently brush the rest off the frame. As you clear a frame, put it into an empty super and cover the box with a heavy, damp cloth to keep the bees out.

Bee escape. People have cleverly designed different mazes and doors that allow bees to leave the supers but keep them from reentering. Place an escape board (or the inner cover fitted with a bee escape) between the super and the hive bodies the day before you plan to harvest honey. At night the bees will leave the supers to hunker down together in the deeps and then will be confounded about how to return. Be sure you insert the escape properly or you will trap bees inside the supers rather than the other way around. I've found it handy to have a bee brush on hand to shoo away any stragglers.

Fume board. There are chemical products available with odors that bees find repugnant. Apply the repellent to the absorbent pad of a fume board (available at a beekeeping supplier or easily made at home). Stack the fume board on top of the super, briefly wait for the bees to run away in horror, and then remove the box. As with any chemical, be sure to follow the directions on the product label exactly.

Gooey, Gluey Dribbles

BY NOW YOU KNOW YOUR BEES and whether they tend to build a lot of burr comb on the top bars. If the boxes usually stick together when you work the hive and honey oozes out as you pull them apart, the day that you remove the super won't be any different. You can avoid bringing a leaking, sticky mess into your house by visiting the hive a day before harvest and scraping off the burr comb. The bees will clean the honey remains up after you leave. Or you can control the mess and bring a plastic garbage bag to layer under the super and catch the drips.

BEE BRUSH

PROS	CONS
Requires only one trip	A lot of bees will end up in the air and may become agitated by the shaking and brushing; with so many bees loose outside the hive, it is harder to keep them from returning to the honey frames
Least expensive option	Works best with fully capped frames; frames with some open cells of unripened honey will get the brush sticky and wet

BEE ESCAPE

PROS	CONS
Stress free for the bees	Requires a visit to the hive a day in advance of harvest
Few bees in the air	If the weather is very hot, the wax may melt if there aren't any bees present to fan it; in this case insert the board at the end of the afternoon and harvest early the next day

FUME BOARD

PROS	CONS
Few bees in the air	Requires the use of chemicals; some brands are classified as hazardous material and smell foul to humans as well as bees, though at least one brand on the market is considered nontoxic and has a pleasant odor
Requires only one trip	Efficacy may depend on sun and heat to evaporate the repellent and disperse the fumes through the super

The Honey Harvest

The honey is now in your house. Before plunging forward, inspect each frame. You want to extract cured honey, not unripe honey or nectar. Nectar won't have the same flavor as cured honey; more importantly, it will have a high moisture content that could cause the honey to ferment if too much nectar is mixed in. All capped cells will hold only cured honey. Open, full cells may contain cured honey that simply hasn't been capped yet, or they may contain nectar or partially ripened honey that you don't want. To test which it is, hold the frame so the comb is parallel to the ground and shake it. Honey is viscous enough to stay in the frame; nectar will spill or drip out.

Even if a frame passes the shake test, set it aside to return it to the hive if more than a third of the cells are uncapped. Without an expensive tool called a refractometer, it's hard to tell if uncapped honey is fully cured or only partially cured, with a moisture level high enough to compromise preservation.

MECHANICAL EXTRACTION

Once you have all the frames that have passed inspection, follow these steps for extracting and packaging the honey.

1. Collect your tools and supplies:
   » Uncapping tool
   » Uncapping tank
   » Extractor
   » Sieve
   » Bucket
   » Bottles
   » Towels and sponges (this is messy work!)

2. Open the honey cells. Use an uncapping tool to remove the wax cappings and an uncapping tank to collect them.

3. After you've uncapped the cells on both sides of a frame, hang it in the tank, which will catch the drips as you continue to work. Repeat until you have enough frames to fill the extractor.

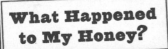

What Happened to My Honey?

AS HONEY AGES it will naturally granulate, a process called crystallization. Some floral nectar, such as goldenrod, produces honey that crystallizes faster than others. The only honey that never crystallizes is from the nectar of tupelo flowers. Crystallized honey is still delicious in spite of its slightly gritty texture. It's not great on toast, but it is perfect in hot tea, which will dissolve the crystals. You can also reliquefy a whole bottle of crystallized honey by standing it in a pot of hot water.

4. Load the extractor and crank it up, sending honey whipping out of the cells with centrifugal force. Be sure the valve on the extractor is open to let the honey pour out and flow through a sieve into the honey bucket. Honey that is warm from the ambient temperature of the room will extract and strain more readily. Repeat until you've processed all the frames. Watch that your bucket doesn't overflow!

Honey drains from the gate on the extractor into a sieve, filtering out slivers of wax and other debris. Through the sieve, the honey flows into a bucket.

5. Drain the uncapping tank into the honey bucket to catch all the drippings from the wax cappings.

6. Cover the honey bucket, and let it sit overnight. Air bubbles and any wax and debris that made it through the sieve will rise to the top and can be skimmed off.

7. You can leave the honey in the bucket indefinitely or bottle it for convenience.

Some people don't strain their honey through a sieve. Chunks of wax will be in the mix, but advocates believe it all adds to the natural goodness.

MANUAL EXTRACTION
Mechanical extractors are the fastest way to collect honey, but it's totally fine to go the old-fashioned route. This method is slow, destroys the comb, and doesn't yield as much honey, but hey, life doesn't have to be all about efficiency.

1. Scrape or cut the comb off the frames.

2. Squeeze the honey out of the wax using your hands, a potato masher, or other tool. If you don't mind getting sticky, using your hands is the simplest option as you can crush the wax directly over a strainer. (Honey is great for the skin!) If you use a tool, mash the wax in a large container such as a pan and then pour the honey into the strainer.

3. Strain the honey using cheesecloth, nylon straining material, a stainless steel window screen, a sieve, a bucket with holes drilled in the bottom, or your own brilliant invention. Have a bucket under the strainer to collect the honey.

4. Cover the bucket, let it sit overnight, and then bottle the honey.

CLEANUP CREW

You now have a colossal, sticky mess of equipment, comb, and wax cappings. What never ceases to amaze me is that those squeaky-clean bees are our allies. Put all the frames in the super and bring it back to the hive. If you expect another honey flow, put the super on the hive as before. If it is the end of the season, stack the super on top of the inner cover. The bees will travel up and clean the honey remains out of the cells, transferring it into the deeps. They don't usually consider any space above the inner cover to be their home and won't put the frames into use, but remove the clean super after a day or two just in case.

Similarly, I spread the wax cappings on a tarp and lay them outside, placing them no closer than twenty feet from the hive so I don't encourage a robbing frenzy. The bees will transform a syrupy muddle into a flaky, dry store of premium beeswax that's good for making into candles or cosmetics. Have you ever had animals that not only cleaned up after themselves but also cleaned up after you?

Our bees are cleaning honey off the mass of sticky cappings we scraped off the frames.

CONGRATULATIONS OR CONDOLENCES

The Cold Hard Facts of Life

I hope you're finding that beekeeping is fun and fascinating. It's also heartbreaking, and you are a very rare beekeeper indeed if you don't lose colonies. Surveys conducted annually by the U.S. Department of Agriculture and Apiary Inspectors of America indicate that national colony losses during the winter alone have hovered between 30 and 35 percent in recent years. The survey doesn't even count the losses that occur during the summer and fall. Also, that number rises about 10 points when you eliminate commercial operations and look just at individual beekeepers' losses.

If your colonies survive, congratulations on your mix of skill and luck. If your colonies die, put your experience into the bigger picture of a high modern rate of colony loss, even among professionals. If it were easy to keep bees alive, we wouldn't be in the midst of a global pollination crisis.

There is a steep learning curve for beginners, and most make mistakes the first few years. Don't get discouraged!

URBAN BEEKEEPING

Bobbi and Greg Marstellar
CHICAGO, ILLINOIS

Bobbi and Greg are rooftop beekeepers who started with a nuc they got from Kress Apiary in Burns Harbor, Indiana. Their mentor there, Bob Kress, has been raising his own local stock for years, and he calls his strain Hoosier bees.

In her blog, Sweet Hive Chicago, Bobbi chronicles the wild and unpredictable joys and heartaches of keeping bees. After her first colony survived a long, bitter winter, including a record-breaking blizzard so severe she dubbed it Snowmageddon, the last bees died in March. She wrote:

> Our hive was morbidly silent. My husband and I crouched in front, desperate for a sign of life. We were bereft and completely mystified. How could our hive have survived the worst part of the winter, then perished in the final days as spring and its promise of pollen were in sight? We thought we'd done everything right. In the end, it wasn't enough. . . . Soon, we will start again with new bees in our hive. I remain an urban beekeeper.

Their mentor, "Bob the Beeman," a seasoned professional who surely did "everything right," lost 80 percent of his hives that year. It's worth remembering that you aren't always to blame for a colony's loss.

Hitting the Skids

Early spring is the most critical time of year. You can't count your chickens until they hatch, and you can't count your hives until nectar and pollen are coming in and the perils of the winter dearth are over. Bees may be dying faster than new bees are born — from disease, mite predation, lack of food, or weather conditions — and the colony may be weakened to a point from which it can't recover.

In late winter or early spring, check your hive to see if anyone is at home. Do you hear any buzzing if you knock on the box?

COMMON REASONS FOR COLONY FAILURE

If there is the awful hush of silence inside the hive, it's time to analyze what happened if possible and clean up the *dead out* (dead colony). Following are some possibilities for the malady responsible.

Starvation. There is a large cluster of dead bees that is not in contact with honey stores. There may be lots of honey in the hive as close as two inches away from the bees, but the cluster was not able to move to reach it (see Winter Cluster, page 123).

Varroa mites. There may be few dead bees in the hive except in regions with very cold winters. Dead bees often have deformed wings or stunted abdomens. The capped brood is scattered, and in the center of what is left of the brood nest, you find small white deposits of mite feces on the ceilings of the cells. There are honey stores left.

Poor ventilation. If condensation built up inside the hive, the bees got wet and died from the chill.

If you see other symptoms, check the appendix on pages 146–150 for more honey bee ailments.

MOVING ONWARD

If your colony died from starvation, mites, or poor ventilation, you can reuse the comb and the honey stores. When the bees die, the mites die, so you won't risk reintroducing Varroa. These resources require careful storage or pests will move in before you get replacement bees. (You are planning to get new bees, right? Of course you are!) Frames that don't have an army of bees to protect and clean the comb and honey are open to infestations of wax moths, hive beetles, and robber bees.

In addition to the recommendations for storing supers on page 119, first scrape any brood off the frames so it doesn't decay. Air out the hive furniture in the sunlight for a week or longer. While outside, turn the boxes on end so they aren't dark, inviting places for critters. Don't put frames with comb outside, since they will attract insects and animals.

Hitting the Jackpot

If you have a colony that survived the winter, congratulations! Once the weather has reliably shifted to spring, there are some maintenance tasks to start you on your second year of beekeeping. In my region, the blooming of dandelions is the sign of a new season.

» Remove winter protections, such as insulation and windbreaks.

» Clean bottom boards.

» Remove entrance reducers if the colony's numbers are strong.

» Feed weak hives. You can do this earlier than spring to save a starving colony. If the weather is still cold, use fondant or a product called winter patties.

The bees probably moved up over the winter into the top box. They may feel cramped and are itching to swarm. If the lower box is mostly empty, relieve congestion by reversing the hive bodies so that the bottom box is now on top and the top one is on the bottom.

And onward you go into the new season! *You are no longer an absolute beginner.*

GLOSSARY

Now that you're practically a veteran and know how to walk the walk, here is a sampling of more advanced jargon so you can also talk the talk.

Absconding. When bees abandon a hive due to extreme stresses

Afterswarms. Additional, smaller swarms that leave the hive with virgin queens, after a hive has swarmed once

AHB. Acronym for Africanized honey bee

Apiculture. The technical word for beekeeping

Balling the queen. When workers form a tight, suffocating cluster around a queen to kill her

Bee gum. A hive located in a hollow log

Beek. Short for beekeeper, a term commonly used in online forums

Brace comb. Comb that bridges two frames, binding the frames together

Brood nest. The area of the hive used for rearing brood

Burr comb. Comb built where the beekeeper doesn't want it, between frames or hive bodies

Chilled brood. Brood that dies from cold because there aren't enough nurse bees to cover the cells and maintain a stable temperature. May be caused by a sudden drop in temperature during rapid spring buildup, by an insufficient number of nurse bees due to a diseased and dwindling colony, or by the beekeeper opening the hive for too long in cool temperatures.

Corbicula. The technical word for a pollen basket

Creamed honey. Honey that has been purposely crystallized into a smooth solid by seeding with 10 percent crystallized honey and storing at about 57°F

Drone congregation area. Where drones assemble to wait for virgin queens to pass

Drone layer. A queen that lays only unfertilized eggs because she is old and has run out of sperm, is poorly mated, is not mated at all, or is sick

Field bees. Another term for forager bees

Honey bound. Describes a colony that has run out of adequate space for the brood nest due to cells filled with honey

Honeydew. A sweet liquid excreted by some insects that eat plant sap, such as aphids; bees may collect and store it as honeydew honey

Hygroscopic. Describes the tendency of honey to absorb water from the air

Hypopharyngeal glands. The glands on a bee's head that produce brood food, including royal jelly

Observation hive. Hive with glass sides so colony can be observed. Used for educational purposes, it usually holds one or two frames.

Orientation flights. The training flights of young bees, usually by large numbers at a time. Orienting bees can be observed hovering in front of the hive.

Parthenogenesis. Birth from unfertilized eggs, as seen with drones

Piping. The noise that queens make, usually when they are newly born. The queens that aren't hatched yet "quack" back, and the new queen seeks and destroys them.

Queen substance. The pheromone that queens secrete from their mandibular glands, which attendants spread around the hive to let the rest of the workers know the colony is queenright. The substance also attracts drones when the queen is a virgin.

Scout bees. Worker bees searching for resources, such as nectar, or a new nesting location for a swarm

Skep. A hive fashioned from plaited rings of straw into a basket. Used for 2,000 years and popularly known as the iconic shape of a hive, skeps are now illegal in the United States. and some other countries, because you can't inspect the hive for disease. Harvesting honey from a skep often meant killing the colony.

Spiracles. The openings on the trachea through which a bee breathes

Splitting (or dividing). Making two or more colonies out of one

Supers. Storage space in the hive for surplus honey

Top bar. The top, horizontal part of a frame, not to be confused with top-bar hives

Travel stains. Darkened areas on the surface of the comb caused by bees walking over it carrying pollen and propolis

Uniting. Making one colony out of two or more. To prevent fighting, a sheet of newspaper is placed between the boxes; the colonies' scents slowly mingle, and the bees chew the paper away.

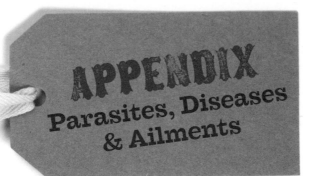

APPENDIX
Parasites, Diseases & Ailments

A strong colony can manage most of the maladies that Mother Nature might throw at it. Bees usually know when something is wrong and make dutiful hygienic efforts to eliminate the presence of parasites, pests, and disease. Stressful conditions, such as low population numbers, a weak queen, a dearth of food, an extended period of bad weather, or overly intrusive management by a beekeeper, can compromise their ability to take care of themselves.

Preventive approaches to take from the start:

» Buy bees and queens from quality suppliers, ideally from local sources that have bees adapted to local conditions.

» Put your hives in a good site and ventilate them adequately. Good sun exposure, especially in the morning, allows the colony to get an early start on foraging and helps suppress many pests and diseases. Avoid sites with cool, damp air.

» Be sure the bees have access to ample healthy food and water.

» Discard comb once it gets dark in color so the bees start fresh with wax that is free of disease organisms and contamination from pesticides.

Finally, don't disturb your bees too much, and with luck you won't see any symptoms of the ailments listed below.

AMERICAN FOULBROOD (AFB)

Because this list is in alphabetical order, the worst comes first. A bacterium that affects a colony's brood, American foulbrood lives up to its nasty name.

Symptoms. The brood cappings appear sunken and discolored, with small holes caused by adult bees that begin to open the cells and then abandon the effort of cleaning out the dead. Brood inside the affected cells is discolored and smells bad, with dead larvae having a melted appearance. For a fairly certain diagnosis, stick a toothpick into a "melted" larva and slowly draw it out. If the larva is ropy and can be drawn into a string an inch long, your colony has American foulbrood.

Cure. Practically impervious to heat, cold, desiccation, and chemical disinfectants, AFB spores can live for more than seventy years. You may have to burn or bury everything — bees, honey, comb, frames, and hive boxes — to prevent its spread. The only deliverance is irradiation, which will also eliminate any other existing pathogens. Some states operate irradiation programs for hive equipment. Check with your closest beekeeping association.

CHALKBROOD

Another brood disease, chalkbrood kills larvae with a covering of fungus that is white or gray. Only colonies that are under stress are susceptible to chalkbrood, though the disease is more prevalent in damp weather.

Symptoms. The larvae look like pieces of hard, white chalk, giving the disease its name, but beekeepers also call affected larvae "mummies." Though symptoms appear after the brood is capped, workers remove the cappings before dragging the mummies out, so you will see a spotty brood pattern.

Cure. The only known cure is a strong population. If wet weather is a contributing factor, move the hive to full sun.

In spite of her lifelike pose, this little one did not survive the winter.

COLONY COLLAPSE DISORDER (CCD)

This disease wipes out colonies on a massive scale and its causes are still poorly understood.

Symptoms. Hives with the following symptoms are diagnosed with CCD:

» Adult workers rapidly disappear from the colony, abandoning the brood.
» No significant quantity of dead bees is found in or near the hive.
» Varroa mites and nosema are not present at a high enough rate to cause colony loss.
» The invasion of pests such as wax moths, hive beetles, and robbing bees, which would ordinarily prey swiftly upon a weak or dead hive, is noticeably delayed.

Cure. There is no definitive, universally agreed-upon cause or cure for this affliction.

DYSENTERY

An intestinal infection caused by amoeba, some pollens, and some honeys, dysentery results in diarrhea.

Symptoms. Brown staining appears in and outside the hive.

Cure. Prevention is the best cure. A sunny location and an upper entrance in cold winter areas will help avoid this ailment. Poor ventilation, routinely damp hives, the consumption of sugar syrup before it's been adequately cured to reduce water content, and prolonged confinement in the hive during cold weather can cause this condition.

EUROPEAN FOULBROOD (EFB)

A bacterial brood disease, European foulbrood is far less serious than American foulbrood. EFB usually appears in colonies that are under stress.

Symptoms. Unlike AFB, affected larvae die when they are just a few days old and the cells are not capped. Though the dead larvae look yellowish and twisted, they aren't ropy. The brood pattern will appear spotty.

Cure. Though colonies used to recover with a strong flow, there are new strains that don't go away as easily. One or two treatments with oxytetracycline will clean the hive.

HIVE BEETLES

These small black beetles lay their eggs in the comb, and when the larvae hatch, they eat wax, pollen, and sometimes eggs and bee larvae. The beetles excrete feces in the honey cells, causing the honey to ferment and become runny.

Symptoms. The telltale sign of an infestation is slimy comb with honey running out.

Cure. As with almost all hive ailments, a strong population of bees is able to defend itself well. To that effect, don't add boxes to the hive until there are enough bees to cover the additional comb. A location in full sun away from woods is best. Space frames tightly and remove crevices or other hiding places for beetles. There are also commercially available solutions.

NOSEMA

Nosema spores invade the digestive tract of adult bees and will aggravate other viruses present in the hive, further weakening the colony. Microscopy is required for diagnosis.

Symptoms. The colony is slow to build up and winter mortality is high. Some affected bees will be too sick to fly.

Cure. The best prevention is a vigorous queen and plenty of pollen or pollen supplement. Chemicals, such as fumigillin, and other products such as Nosevit and Apiherb are available for treatment. If the colony dies from nosema, freeze or expose the frames to sunlight.

PESTICIDES

Agricultural pesticides that are overtly lethal to bees usually kill individual foragers before they return to the hive; however, chemicals are known to be responsible for some sudden colony loss.

Researchers are studying sublethal effects of pesticides: those that don't kill bees or colonies immediately but compromise the health of a colony significantly enough that it collapses later. Toxic effects include decreased longevity in adult bees, weakened immune systems, impairment of memory and learning, and damage to developing brood. There is also evidence that interactions between chemicals — a synergy between fungicides and miticides, for example — can cause a level of toxicity that is greater than the sum of its parts.

Symptoms. A sudden colony loss is often characterized by large numbers of bees piled outside the hive. Though every beekeeper should be concerned about pesticide use, those who have their hives situated near large farms using intensive chemicals have the most serious problem.

Cure. If you use pesticides in the hive at all, use them only to manage a clear and present danger rather than as a preventive measure. Avoid pesticides on your own property. If you are experiencing colony loss due to pesticide use in your area, scout another location. Farmers growing organic crops often welcome beekeepers to site hives on their property.

Even chemicals that beekeepers use to treat ailments in the hive often interfere with a colony's natural immunological defenses. We cure one malady only to open the door to other afflictions.

TRACHEAL MITES

These mites are bloodsuckers, like Varroa, but live in a bee's breathing tube. The mites shorten a bee's life span, inhibit her ability to cluster well in cold weather, and transmit a virus that deforms her wings in a disjointed configuration known as *K* wings.

Symptoms. When a colony is dying from an infestation, large numbers of bees will leave the hive, unable to fly. Individual bees have deformed K wings. (The body forms the vertical stroke of the K, and the wings, the diagonal strokes.) The hive may not survive the winter, in which case dead bees will be scattered in the hive in small clusters instead of one large cluster. *Note:* The virus that results in K wings may be present independently of tracheal mites but is not a significant threat to the colony's survival without the added stress of the parasitic mites.

Cure. Most stocks are resistant to tracheal mites, so this pest is not a widespread concern. As needed, there are commercial medications available, such as menthol or formic acid fumigation.

VARROA MITES
See a full discussion of Varroa mites beginning on page 109.

VIRUSES
Colonies will recover from most virus infections if mite levels are low, there is a vigorous queen, and the colony has access to good nutrition. Following are some viruses you might see.

Sacbrood. After the cell is capped, the larvae become discolored and look like sacs filled with watery liquid.

Chronic Paralysis Virus. Black, shiny bees with tattered wings will look like they are trembling.

Deformed Wing Virus. DWV is usually associated with Varroa mites and is the number one colony killer. Adult bees that are infected will have shriveled wings. Late larvae and pupae turn gray and die.

WAX MOTHS
Adult moths lay eggs in the comb, and larvae will chew through the wax as they move through the hive. The destruction of the wax is collateral damage as the moth larvae root around to find pollen, bee cocoons, and cast bee larvae skins. Consequently, they attack older comb, rather than freshly drawn wax.

Symptoms. The telltale signs of a wax moth infestation are tough cocoons affixed to the top bars and boxes and a mass of webbing that surfaces the frames. The larvae spin these webs, called *galleries*, for protection from the bees.

Cure. A healthy colony is able to remove the moth larvae before they do any damage, but a weak or dead hive is a sitting duck. Wax moths are a considerable nuisance when storing frames unless the equipment is protected (see page 120).

RESOURCES

Books and References

BEESOURCE.COM
www.beesource.com
An active online beekeeping community, Bee Source hosts forums in which you can read posts from thousands of beekeepers sharing their own knowledge, opinions, and experiences. If you register as a member, you can contribute your own questions and answers. There are also articles and resources, and all information and services are free.

BUSH, MICHAEL. *The Practical Beekeeper,* Vol. 1. X-Star Publishing Company, 2004.
Michael Bush is a trailblazer in natural beekeeping. He has kept bees for almost forty years, and since 2000 has been experimenting with methods for sustaining honey bee health without treating for pests and diseases. In an unassuming and engaging style, he recounts what he has found successful in three self-published volumes titled *The Practical Beekeeper,* the first volume of which is for beginners. He publishes all of the information in those books for free on his website www.bushfarms.com.

FISHER, ROSE-LYNN. *Bee.* Princeton Architectural Press, 2012.
Fisher presents sixty artful photographs, taken with an electron microscope, that magnify the honey bee hundreds to thousands of times. There are incredible discoveries, such as a bee's complex eyes comprising 6,900 hexagonal lenses, resembling the structure of comb.

FLOTTUM, KIM. *The Backyard Beekeeper,* revised edition. Quarry Books, 2010.
As the editor of *Bee Culture* magazine and coeditor of the 41st edition of *The ABC & XYZ of Bee Culture,* the bible of U.S. beekeeping, the author has both feet on the terra firma of mainstream beekeeping. His book is a good overview of the basics.

LIVING WITH WILDLIFE FOUNDATION
www.lwwf.org
This nonprofit organization offers a free PDF download with thorough information on electric fencing.

LOVELL, JOHN. *Honey Plants of North America.* A.I. Root Company, 1999.
Originally published in 1926, this book hasn't been updated other than the release of a paperback edition in 1999. However, there is currently no better source for information about which plants are sources of nectar, where they grow, what they look like, when they bloom, and what kind of honey they produce.

THE MID-ATLANTIC APICULTURAL RESEARCH AND EXTENSION CONSORTIUM. *A Field Guide to Honey Bees and Their Maladies.* AG Communications and Marketing, 2011.

A visual guide of parasites and diseases, this book is also available as a free download at http://extension.psu.edu/start-farming/bees

READICKER-HENDERSON, E. *A Short History of the Honey Bee.* Timber Press, 2009.

Take a break from reading how-to information on keeping bees and pick up this book to remember why bees fascinated you in the first place. Looking at bee behavior and beekeeping practices through history, the author writes in a prose style that is poetic rather than practical.

ROOT, AMOS IVES, ANN HARMAN, H. SHIMANUKI, AND KIM FLOTTUM. *The ABC & XYZ of Bee Culture,* 41st edition. A.I. Root Company, 2007.

Though not necessarily for the beginner at an exhaustive 911 pages, no resource list would be complete without this foundational encyclopedia.

SAMMATARO, DIANA AND ALPHONSE AVITABILE. *The Beekeeper's Handbook,* 4th edition. Cornell University Press, 2011.

Well-organized, clearly and succinctly written, and thorough without being exhaustingly comprehensive, this book is an excellent addition to your reference library. With a pedagogical tone and minimally illustrated in black and white, it won't be the friendliest or most accessible guide you own, but will come in handy when you want to research a particular topic.

SCIENTIFIC BEEKEEPING
www.scientificbeekeeping.com
The author of this site, Randy Oliver, provides summaries of current scientific research to help beekeepers plan their management strategies for colony health. He also offers observations and data based upon his own experience and field research while managing a 1,000-colony migratory beekeeping operation in California. He balances a no-nonsense commitment to hard facts with an open-minded interest in progressive ideas.

STIGLITZ, DEAN AND LAURIE HERBOLDSHEIMER. *The Complete Idiot's Guide to Beekeeping.* Alpha Books, 2010.
This excellent book advocates small-cell beekeeping and other approaches for treatment-free management of colonies. Includes detailed information on regressing bees raised on standard-cell comb to a natural, smaller size.

TOURNERET, ÉRIC. *Le Peuple des Abeilles.* France: Rustica, 2007.
An art book of the most amazing bee photos I've ever had the privilege to see. Only available as an import, the book is pricey but worth every penny and can be purchased at www.abebooks.com. If you are pinching pennies at the moment, enjoy many of the photos online at http://thehoneygatherers.com.

USDA BELTSVILLE AGRICULTURAL RESEARCH CENTER
www.ba.ars.usda.gov/psi/brl
If your colony fails, you can send some of the dead bees to this research laboratory and staff will perform a sort of autopsy for you, identifying the pests and diseases present. Visit the website for instructions about how to submit samples. This service is free of charge.

THE XERCES SOCIETY. *Attracting Native Pollinators.* Storey Publishing, 2011.
If you are interested in bees for local pollination, you need this book. One solution is to look beyond the honey bee. The Xerces Society provides detailed plans for providing flowering habitats and nesting sites to attract pollinating bees, wasps, butterflies, moths, flies, and beetles that are native to your region.

Suppliers

BEE THINKING
503-770-0233
www.beethinking.com
Both online and in their bricks-and-mortar store in Portland, Oregon, Matt and Jill Reed sell foundation-less top-bar and Warré hives of their own design that are manufactured in a local mill from sustainably harvested Western Red Cedar.

BRUSHY MOUNTAIN BEE FARM
800-233-7929
www.brushymountainbeefarm.com
Brushy Mountain is a good, reliable source for all your general beekeeping supplies.

DADANT & SONS, INC.
888-922-1293
www.dadant.com
Dadant advertises itself as the oldest and largest manufacturer of beekeeping supplies and is an excellent, well-respected resource.

MANN LAKE LTD.
800-880-7694
www.mannlakeltd.com
While small-cell wax foundation is common at many suppliers, if you want to start with small-cell plastic one-piece frames, Mann Lake is your source, along with all your other beekeeping needs.

Contributors

Visit some of the beekeepers profiled in this book on their own turf.

AMY AZZARITO
Design Sponge
www.designsponge.com/author/amya

BACKWARDS BEEKEEPERS
www.backwardsbeekeepers.com

BOBBI AND GREG MARSTELLAR
Sweet Hive Chicago
www.sweethivechicago.com

CHANTAL FORSTER
Mistress Beek
www.mistressbeek.com

KARL ARCURI
Urban Beekeeping in Austin, Texas
www.austinbeekeeping.com

MATT AND JILL REED
Bee Thinking
www.beethinking.com

STEVEN CAMERON
San Francisco Bee-Cause
www.sfbeecause.org

INDEX

ADDITIONAL INTERIOR PHOTOGRAPHY CREDITS

Now that you've mastered the first year of raising honey bees, it's time to explore chickens, goats, and more with these other Storey titles!

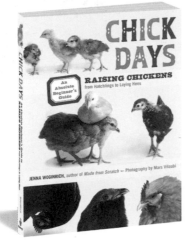

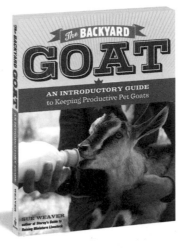

CHICK DAYS by Jenna Woginrich
Photography by Mars Vilaubi
128 pages. Paper. ISBN 978-1-60342-584-1.

THE BACKYARD GOAT by Sue Weaver
224 pages. Paper. ISBN 978-1-60342-790-6.

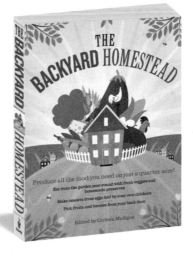

STOREY'S GUIDE TO KEEPING HONEY BEES
by Malcom T. Sanford and Richard E. Bonney
256 pages. Paper. ISBN 978-01-60342-550-6.

THE BACKYARD HOMESTEAD
edited by Carleen Madigan
368 pages. Paper. ISBN 978-1-60342-138-6.

These and other books from Storey Publishing are available
wherever quality books are sold or by calling 1-800-441-5700.
Visit us at *www.storey.com*.